Lecture Notes in Mathematics 2152

More information about this series at http://www.springer.com/series/304

David Eisenbud • Irena Peeva

Minimal Free Resolutions over Complete Intersections

David Eisenbud
MSRI & Mathematics Department
University of California
Berkeley, CA, USA

Irena Peeva
Mathematics Department
Cornell University
Ithaca, NY, USA

ISSN 0075-8434 ISSN 1617-9692 (electronic)
Lecture Notes in Mathematics
ISBN 978-3-319-26436-3 ISBN 978-3-319-26437-0 (eBook)
DOI 10.1007/978-3-319-26437-0

Library of Congress Control Number: 2015958653

Mathematics Subject Classification (2010): 13D02, 11-XX

Springer Cham Heidelberg New York Dordrecht London

Printed on acid-free paper

Springer International Publishing AG Switzerland is part of Springer Science+Business Media (www.springer.com)

Preface

The theory of *higher matrix factorizations of a regular sequence* $f_1, \dots, f_c$ presented in this book is an extension of the theory of matrix factorizations of an element in a commutative ring. It allows us to give a description of the eventual structure of minimal free resolutions over a complete intersection $R = S/(f_1, \dots, f_c)$ of arbitrary codimension c (where S is a regular local ring). We construct the minimal free resolutions of high R-syzygy modules, both over the regular local ring S and over the complete intersection ring R.

Berkeley, CA, USA
Ithaca, NY, USA
August 2015

David Eisenbud
Irena Peeva

Acknowledgments

We are grateful to Jesse Burke and Hailong Dao who read a draft of the manuscript and made helpful comments. We want to express our thanks to David Morrison for lecturing to us on the applications of matrix factorizations in physics. We thank Luchezar Avramov and Ragnar Buchweitz for useful conversations. We profited from examples computed in the system Macaulay2 [M2], and we wish to express our thanks to its programmers Mike Stillman and Dan Grayson.

Eisenbud is partially supported by NSF grant DMS-1001867, and Peeva is partially supported by NSF grants DMS-1100046 and DMS-1406062 and by a Simons Fellowship; both authors are partially supported by NSF grant 0932078000 while in residence at MSRI.

Contents

Chapter 1
Introduction and Survey

Abstract We begin the chapter with some history of the results that form the background of this book. We then define higher matrix factorizations, our main focus. While classical matrix factorizations are factorizations of a single element, higher matrix factorizations deal directly with sequences of elements. In Sect. 1.3, we outline our main results. Throughout the book, we use the notation introduced in Sect. 1.4.

1.1 How We Got Here

Since the days of Cayley [16] and Hilbert [32], minimal free resolutions of finitely generated modules have played many important roles in mathematics. They now appear in fields as diverse as algebraic geometry, invariant theory, commutative algebra, number theory, and topology.

Hilbert showed that the minimal free resolution of any module over a polynomial ring is finite. In the hands of Auslander, Buchsbaum and Serre this property was identified with the geometric property of non-singularity (which had been identified algebraically by Zariski in [58]): local rings for which the minimal free resolution of every module is finite are the *regular local rings*. The Auslander-Buchsbaum formula also identifies the length of the minimal free resolution of a module as complementary to another important invariant, the depth.

The minimal free resolution of the residue field $k = S/\mathbf{m}$, where $\mathbf{m}$ denotes the maximal ideal of a local ring S, plays a special role: in the case of a regular ring this is the Koszul complex (the name is standard, though the idea and the construction were present in the work of Cayley, a century before Koszul).

The condition of minimality is important in these theories. The mere existence of free resolutions suffices for foundational issues such as the definition of Ext and Tor, and there are various methods of producing resolutions uniformly (for example the Bar resolution, in the case of algebras over a field). But without minimality, resolutions are not unique, and the very uniformity of constructions like the Bar resolution implies that they give little insight into the structure of the modules resolved. By contrast, the minimal resolution of a finitely generated module over a local ring is unique and contains a host of subtle invariants of the module. There

D. Eisenbud, I. Peeva, *Minimal Free Resolutions over Complete Intersections*,
Lecture Notes in Mathematics 2152, DOI 10.1007/978-3-319-26437-0_1

are still many mysteries about minimal free resolutions over regular local rings, and this is an active field of research.

Infinite minimal free resolutions seem first to have come to the fore around 1957 in the work of Tate [57], perhaps motivated by constructions in group cohomology over a field of characteristic $p > 0$ coming from class field theory. The simplest interesting case is that of the group algebra $k[G]$ where G is a group of the form

$$G = \mathbb{Z}/(p^{a_1}) \oplus \cdots \oplus \mathbb{Z}/(p^{a_1})$$

and k is a field of characteristic p. In this case $k[G]$ is a local ring of the special form

$$k[G] \cong k[x_1, \dots, x_c]/(f_1, \dots, f_c).$$

where $f_i(x) = x_i^{p^{a_i}} - 1$ for each i, and the maximal ideal is generated by $(x_1 - 1, \dots, x_c - 1)$. The cohomology of such a group, with coefficients in a $k[G]$-module N is, by definition, $\mathrm{Ext}_{k[G]}(k, N)$, and it is thus governed by a free resolution of the residue field k as a module over $k[G]$. The case of non-commutative groups is of course the one of primary interest; but it turns out that many features of resolutions over non-commutative group algebras are governed by the resolutions over the elementary abelian p-groups related to them, so the commutative case plays a major role in the theory.

Generalizing the example of the group algebras above, Tate gave an elegant description of the minimal free resolution of the residue field k of a ring R of the form

$$R = S/(f_1, \dots, f_c)\,,$$

where S is a regular local ring with residue field k and $f_1, \dots, f_c$ is a *regular sequence*—that is, each f_i is a non-zerodivisor modulo the ideal generated by the preceding ones and $(f_1, \dots, f_c) \neq S$. Such rings are usually called *complete intersections*, the name coming from their role in algebraic geometry. It is with minimal free resolutions of arbitrary finitely generated modules over complete intersections that this book is concerned.

Tate showed that, if R is a complete intersection, then the minimal free resolution of k has a simple structure, just one step removed from that of the Koszul complex. Tate's paper led to a large body of work about the minimal free resolutions of the residue fields of other classes of rings (see for example the surveys by Avramov [5] and McCullough-Peeva [41]). Strong results were achieved for complete intersections, but the structures that emerged in more general cases were far more complex. Still today, our knowledge of infinite minimal free resolutions of modules, or even of the residue class field, over rings that are not regular or members of a few other special classes (Koszul rings, Golod rings) is very slight.

The situation with complete intersections has drawn a lot of interest. Shamash [55] (and later, in a different way, Eisenbud [25]) showed how to generalize Tate's construction to any module ... but, except for the residue class field, this method rarely produces a minimal resolution. A different path was begun in the 1974 paper [31] of Gulliksen, who showed that if N and P are any finitely generated modules over a complete intersection, then $\mathrm{Ext}_R(N,P)$ has a natural structure of a finitely generated graded module over a polynomial ring $\mathcal{R} = k[\chi_1,\ldots,\chi_c]$, where c is the codimension of R. He used this to show that the Poincaré series $\sum_i \beta_i^R(N)x^i$, the generating function of the Betti numbers $\beta_i^R(N)$, is rational and that the denominator divides $(1-x^2)^c$. In 1989 Avramov [4] identified the dimension of $\mathrm{Ext}_R(N,k)$, which he called the *complexity* of N, with a correction term in a natural generalization of the Auslander-Buchbaum formula.

Gulliksen's finite generation result implies that the even Betti numbers $\beta_{2i}^R(N)$ are eventually given by a polynomial in i, and similarly for the odd Betti numbers. Avramov [4] proved that the two polynomials have the same degree and leading coefficient. In 1997 Avramov, Gasharov and Peeva [7] gave further restrictions on the Betti numbers, establishing in particular that the Betti sequence $\{\beta_i^R(N)\}$ is eventually either strictly increasing or constant.

The paper [25] of Eisenbud in 1980 brought a somewhat different direction to the field. He took the point of view that what is simple about the minimal free resolution

$$\cdots \longrightarrow F_n \longrightarrow \cdots \longrightarrow F_1 \longrightarrow F_0$$

of a module $N = \mathrm{Coker}(F_1 \longrightarrow F_0)$ over a regular ring S is that the F_i are *eventually* 0 (by Hilbert's Syzygy Theorem): in fact they are 0 for all $i > \dim(S)$ (this would not be true if we did not insist on *minimal* resolutions).

Eisenbud described the eventual behavior of minimal free resolutions of arbitrary finitely generated modules over *hypersurface rings*. These are the rings of the form $R = S/(f)$ where S is a regular local ring, that is, complete intersections of codimension 1. He proved that, in this case, the minimal resolution of every finitely generated R-module eventually becomes periodic, of period at most 2, and that these periodic resolutions correspond to *matrix factorizations* of the defining equation f; that is, to pairs of square matrices (A,B) of the same size such that $AB = BA = f \cdot I$, where I is an identity matrix. As a familiar example, if $f = \det(A)$ then we could take B to be the adjoint matrix of A. As with the case of regular rings, the simple pattern of the matrix factorization starts already after at most $\dim(R)$ steps in the resolution. The theory gives a complete and powerful description of the eventual behavior of minimal resolutions over a hypersurface.

Since there exist infinite minimal resolutions over any singular ring, and the hypersurface ring R is singular if and only if $f \in \mathbf{m}^2$ (where $\mathbf{m}$ is the maximal ideal of S), it follows that matrix factorizations exist for any element $f \in \mathbf{m}^2$—for instance, for any power series of order ≥ 2. Such a power series f has an ordinary factorization—that is, a factorization by 1×1 matrices—if and only if f defines a reducible hypersurface. Factorization by larger matrices seems first to have appeared in the work of P.A.M Dirac, who used it to find a matrix square root—now called the *Dirac operator*—of an irreducible polynomial in partial derivatives

$$\frac{1}{c^2}\frac{\partial^2}{\partial t^2} - \frac{\partial^2}{\partial x^2} - \frac{\partial^2}{\partial y^2} - \frac{\partial^2}{\partial z^2}.$$

The general theory of matrix factorizations is briefly described in Chap. 2. There are still many open questions there, such as: "*What determines the minimal size of the matrices in the factorizations of a given power series?*"

Matrix factorizations have had many applications. Starting with Kapustin and Li [37], who followed an idea of Kontsevich, physicists discovered amazing connections with string theory—see [1] for a survey. A major advance was made by Orlov [43, 44, 46, 47], who showed that matrix factorizations could be used to study Kontsevich's homological mirror symmetry by giving a new description of singularity categories. Matrix factorizations have also proven useful for the study of cluster tilting [18], Cohen-Macaulay modules and singularity theory [9, 12, 15, 40], Hodge theory [11], Khovanov-Rozansky homology [38, 39], moduli of curves [51], quiver and group representations [2, 4, 36, 52], and other topics, for example, [10, 17, 21–23, 33–35, 50, 53–56].

What about the eventual behavior of minimal free resolutions over other rings? The Auslander-Buchsbaum-Serre Theorem shows that regular rings are characterized by saying that minimal free resolutions are eventually zero, suggesting that the eventual behavior of resolutions over a non-regular ring R corresponds to some feature of the singularity of R. This idea has been extensively developed by Buchweitz, Orlov and others under the name "singularity category"; see [43–47].

The obvious "next" case to study after hypersurface rings is the eventual behavior of minimal resolutions over complete intersections.

One useful method of extending the theory of matrix factorizations to complete intersections was developed by Orlov [45] and subsequent authors, for example [13, 14, 51]. This method regards a complete intersection as a family of hypersurfaces parametrized by a projective space. For example, suppose that $S = k[x_1, \ldots, x_n]$ is the coordinate ring of the affine n-space $\mathbb{A}^n_k$ over a field k, and R is the complete intersection $R = S/(f_1, \ldots, f_c)$. Consider the element $f = \sum_1^c z_i f_i \in S[z_1, \ldots, z_c]$ as defining a hypersurface in the product of $\mathbb{A}^n$ and the projective space $\mathbb{P}^{c-1}$ and consider the category of matrix factorizations (now defined as a pair of maps of vector bundles) of f. This idea has proven useful in string theory and elsewhere, and provides a good definition of a singularity category for complete intersections; but it does not seem to shed any light on the structure of minimal free resolutions over R.

Eisenbud provided easy examples showing that over complete intersections, unlike regular and hypersurface rings, nice patterns may begin only far back in the minimal free resolutions. Even though Eisenbud's paper is entitled "Homological Algebra over a Complete Intersection", it only gives strong results for the case of hypersurfaces and really doesn't get much further than posing questions about minimal free resolutions over complete intersections of higher codimension. In the case of codimension 2, important steps were taken by Avramov and Buchweitz in [6] in 2000 using the classification of modules over the exterior algebra on two variables; in particular they constructed minimal free resolutions of high syzygies over a codimension two complete intersection as quotients. But the general case (of higher codimensions) has remained elusive.

Nevertheless, when one looks at the matrices of the differential in minimal resolutions over a complete intersection, for example in the output of the computer systems Macaulay2 [30] and Singular [19], one feels the presence of repetitive patterns.

The authors of this book have wondered, for many years, how to describe the eventual patterns in the minimal resolutions of modules over complete intersections of higher codimension. With the theory presented here we believe we have found an answer: when M is a sufficiently high syzygy over a complete intersection ring R, our theory describes the minimal free resolutions of M as an S module and as an R-module.

Revealing the Pattern

In the next section we will introduce the notion of a Higher Matrix Factorization. Ordinary matrix factorizations allow one to understand minimal free resolutions of high syzygies over a hypersurface ring in terms of a simple matrix equation; they show in particular that such resolutions are eventually periodic. We introduce higher matrix factorizations to give, in an equational form, the data needed to describe the structure of minimal free resolutions of high syzygies over a complete intersection of arbitrary codimension; like ordinary matrix factorizations, they show that minimal resolutions eventually exhibit stable patterns.

We define higher matrix factorizations in the next section. Here we provide motivation, by sketching how higher matrix factorizations arise in the structure of minimal free resolutions of high syzygies:

Let N be any module over a complete intersection

$$R = S/(f_1, \dots, f_c),$$

where S is a regular local ring and $f_1, \dots, f_c$ is a regular sequence, and let M be a sufficiently high syzygy—a *stable syzygy* in the sense of Chap. 6—over R. The module M is, in particular, a maximal Cohen-Macaulay R-module without free summands. We begin with replacing the sequence $f_1, \dots, f_c$ by a generic choice of generators of the ideal $(f_1, \dots, f_c)$. We will analyze the minimal free resolution of M over R by an induction on the codimension c. The case $c = 0$ is the case of the regular local ring S: the only stable syzygy over S is the module 0.

We thus assume, by induction, that the minimal resolutions of stable syzygies over the ring $R' = S/(f_1, \dots, f_{c-1})$ are understood. As mentioned above, the Shamash construction rarely produces minimal resolutions; but (because of our general position hypothesis on the generators $f_1, \dots, f_c$ and the definition of a stable syzygy) the minimal free resolution of M over R is actually obtained as the Shamash construction starting from a minimal free resolution $\mathbf{U}$ of M as an R' module; thus it suffices to obtain such a resolution. Of course M is *not* a stable syzygy over R'—it is not even a maximal Cohen-Macaulay R' module. One of the main new ideas of this book is that $\mathbf{U}$ can be constructed in a simple way, which we call a *Box complex*, described as follows.

Like any maximal Cohen-Macaulay module over a Gorenstein ring, M is the second syzygy over R of a unique maximal Cohen-Macaulay module L without free summands. We define $M' := \mathrm{Syz}_2^{R'}(L)$ to be the second syzygy of L as a module over R' or, equivalently, as the non-free part of the Cohen-Macaulay approximation of M over R'. We prove that M' is a stable syzygy over R', and thus has a minimal free resolution described by a higher matrix factorization—this will be part of the higher matrix factorization of M.

Let

$$\cdots \xrightarrow{\partial} A_2' \xrightarrow{\partial} A_1' \xrightarrow{\partial} A_0'$$

be the minimal free resolution of M' over R', and let

$$\overline{b} : \overline{B}_1 \longrightarrow \overline{B}_0 \longrightarrow L \longrightarrow 0$$

be the minimal free presentation of L as an R-module. We prove that any lifting $b' : B_1' \longrightarrow B_0'$ of b to a map of free R'-modules gives a minimal free presentation of L as an R'-module, and hence the minimal free resolution of L as an R'-module has the form

$$\cdots \xrightarrow{\partial} A_2' \xrightarrow{\partial} A_1' \xrightarrow{\partial} A_0' \longrightarrow B_1' \xrightarrow{b'} B_0'.$$

Since L is annihilated by f_c there is a homotopy for f_c on this complex. Let ψ' denote the component of that homotopy that maps $B_1' \longrightarrow A_0'$. We prove that the minimal free resolution of M over R' is

$$\begin{array}{ccccccc}
\cdots \to A_3' & \xrightarrow{\partial'} & A_2' & \xrightarrow{\partial'} & A_1' & \xrightarrow{\partial'} & A_0' \\
 & & & & \oplus & \nearrow_{\psi'} & \oplus \\
 & & & & B_1' & \xrightarrow{b'} & B_0',
\end{array}$$

which we call the *Box complex*. We show that the minimal free resolution of M over R is the Shamash construction applied to this Box complex.

We will define a Higher Matrix Factorization for M in the next section in a way that captures the information in the "box":

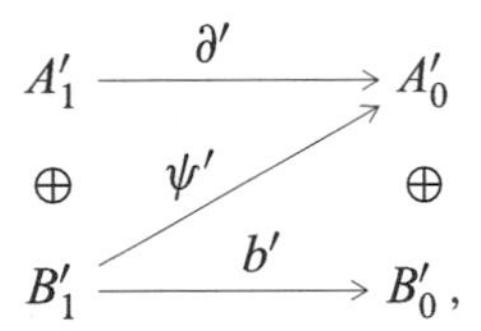

where A'_1 and A'_0 come inductively from a higher matrix factorization for $f_1, \ldots, f_{c-1}$.

The data in the higher matrix factorization suffice to describe the minimal resolutions of the stable R-syzygy M both as an S-module and as an R-module. These constructions and some consequences are outlined in Sect. 1.3.

1.2 What is a Higher Matrix Factorization?

The main concept we introduce in this book is that of a *higher matrix factorization* (which we sometimes abbreviate HMF) with respect to a sequence of elements in a commutative ring. To emphasize the way in which it generalizes the classical case, we first briefly recall the definition, from [25], of a *matrix factorization* with respect to a single element. More about this case can be found in Chap. 2.

Definition 1.2.1 If $0 \neq f \in S$ is an element in a commutative local ring then a *matrix factorization* with respect to f is a pair of finitely generated free modules A_0, A_1 and a pair of maps

$$A_0 \xrightarrow{h} A_1 \xrightarrow{d} A_0$$

such that the diagram

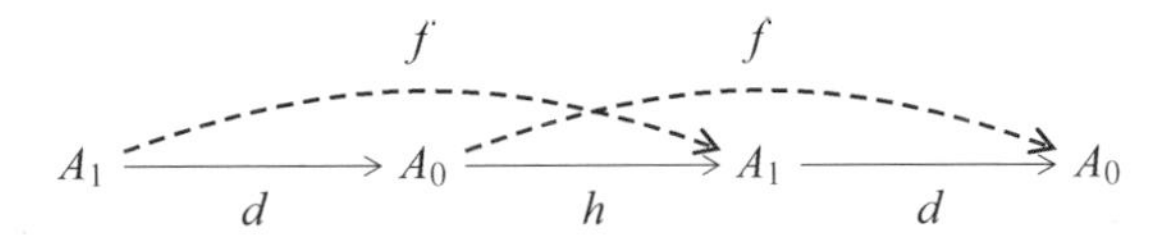

commutes or, equivalently:

$$dh = f \cdot \mathrm{Id}_{A_0} \quad \text{and} \quad hd = f \cdot \mathrm{Id}_{A_1} \,. \tag{1.1}$$

If f is a non-zerodivisor, S is local with residue field k, and both d and h are *minimal* maps in the sense that $d \otimes k = 0 = h \otimes k$, then the matrix factorization allows us to describe the minimal free resolutions of the module $M := \mathrm{Coker}(d)$

over the rings S and $R := S/(f)$; they are:

$$0 \longrightarrow A_1 \xrightarrow{d} A_0 \longrightarrow M \longrightarrow 0 \text{ over } S; \text{ and}$$

$$\cdots \xrightarrow{R\otimes d} R \otimes A_0 \xrightarrow{R\otimes h} R \otimes A_1 \xrightarrow{R\otimes d} R \otimes A_0 \longrightarrow M \longrightarrow 0 \text{ over } R. \quad (1.2)$$

By [25], minimal free resolutions of all sufficiently high syzygies over a hypersurface ring have this form.

Matrix Factorizations of a Sequence of Elements

Definition 1.2.2 Let $f_1, \ldots, f_c \in S$ be a sequence of elements of a commutative ring. A *higher matrix factorization* (d, h) *with respect to* $f_1, \ldots, f_c$ is:

(1) A pair of finitely generated free S-modules A_0, A_1 with filtrations

$$0 \subseteq A_s(1) \subseteq \cdots \subseteq A_s(c) = A_s, \quad \text{for } s = 0, 1,$$

such that each $A_s(p-1)$ is a free summand of $A_s(p)$;

(2) A pair of maps d, h preserving filtrations,

$$\bigoplus_{q=1}^{c} A_0(q) \xrightarrow{h} A_1 \xrightarrow{d} A_0,$$

where we regard $\oplus_q A_0(q)$ as filtered by the submodules $\oplus_{q \le p} A_0(q)$;

such that, writing

$$A_0(p) \xrightarrow{h_p} A_1(p) \xrightarrow{d_p} A_0(p)$$

for the induced maps, the diagrams

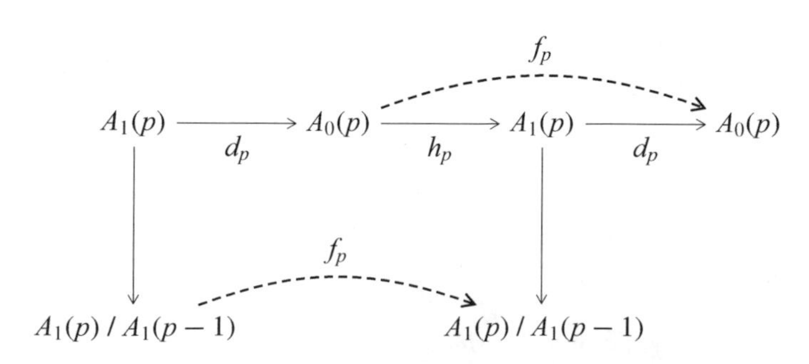

commute modulo $(f_1, \dots, f_{p-1})$ for all p; or, equivalently,

(a) $d_p h_p \equiv f_p \operatorname{Id}_{A_0(p)} \bmod (f_1, \dots, f_{p-1})A_0(p)$;
(b) $\pi_p h_p d_p \equiv f_p \pi_p \bmod (f_1, \dots, f_{p-1})\big(A_1(p)/A_1(p-1)\big)$, where π_p denotes the projection $A_1(p) \longrightarrow A_1(p)/A_1(p-1)$.

Set $R = S/(f_1, \dots, f_c)$. We define *the module of the higher matrix factorization* (d, h) to be

$$M := \operatorname{Coker}(R \otimes d)\,.$$

We refer to modules of this form as *higher matrix factorization modules* or *HMF modules*.

If S is local, then we call the higher matrix factorization *minimal* if d and h are minimal maps (that is, the image of each map is contained in the maximal ideal times the target).

In Sect. 8.1, we show that a homomorphism of HMF modules induces a morphism of the whole higher matrix factorization structure; see Definition 8.1.1 and Theorem 8.1.2 for details.

For each $1 \le p \le c$, we have a higher matrix factorization $(d_p, (h_1|\cdots|h_p))$ with respect to $f_1, \dots, f_p$, where $(h_1|\cdots|h_p))$ denotes the concatenation of the matrices $h_1, \dots, h_p$ and thus an HMF module

$$M(p) = \operatorname{Coker}(S/(f_1, \dots, f_p) \otimes d_p)\,.$$

This allows us to do induction on p. We will show in Theorem 7.4.1 that if S is Gorenstein then the modules $M(p)$ arise as the essential Cohen-Macaulay approximations of M over the rings $R(p) = S/(f_1, \dots, f_p)$, and on the other hand they arise as syzygies over the rings $R(p)$ of a single R-module.

Definition 1.2.3 Let (d, h) be a higher matrix factorization. Use the notation in Definition 1.2.2 and choose splittings so that

$$A_s(p) = \bigoplus_{q=1}^{p} B_0(q)$$

for all p and $s = 0, 1$. We say that (d, h) is a *strong* matrix factorization if it satisfies

$$(\mathrm{a}') \qquad d_p h_p \equiv f_p \operatorname{Id}_{A_0(p)} \bmod \Bigg(\sum_{1 \le r < q \le p} f_r B_0(q) \Bigg)\,,$$

which is a stronger condition than condition (a). We shall see in Sect. 5.3 that this property holds if and only if it is possible to extend h to a homotopy on the S-free resolution of the HMF module M constructed in Chap. 3. We will show in Theorem 5.3.1 that if (d, h) is a higher matrix factorization for a regular sequence

$f_1, \dots, f_c$, then there exists a strong matrix factorization (d, g) with the same HMF module $M = \text{Coker}(R \otimes d)$.

Example 1.2.4 Let $S = k[a, b, x, y]$ over a field k, and consider the complete intersection $R = S/(xa, yb)$. Let $N = R/(x, y)$. The module N is a maximal Cohen-Macaulay R-module. The earliest syzygy of N that is an HMF module is the third syzygy M. We can describe the higher matrix factorization for M as follows. After choosing a splitting

$$A_s(2) = A_s(1) \oplus B_s(2),$$

we can represent the map d as

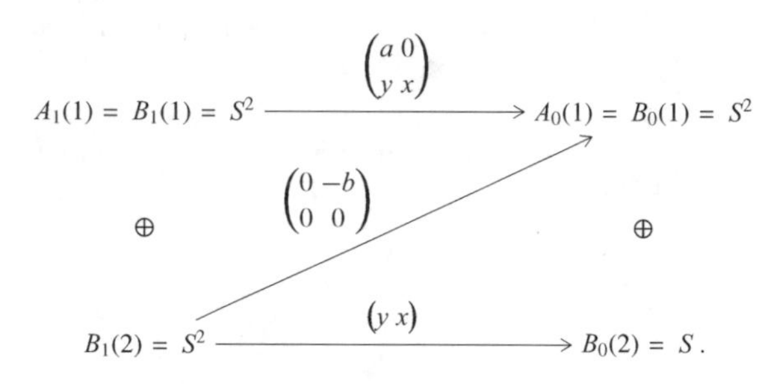

The pair of maps

$$d_1 : A_1(1) \xrightarrow{\begin{pmatrix} a & 0 \\ y & x \end{pmatrix}} A_0(1) \quad \text{and} \quad h_1 : A_0(1) \xrightarrow{\begin{pmatrix} x & 0 \\ -y & a \end{pmatrix}} A_1(1)$$

forms a matrix factorization for the element xa since $d_1 h_1 = h_1 d_1 = xa\,\text{Id}$. The maps

$$h_2 : A_0 = A_0(2) \longrightarrow A_1 = A_1(2)$$

and

$$d_2 : A_1 = A_1(2) \longrightarrow A_0 = A_0(2)$$

are given by the matrices

$$h_2 = \begin{pmatrix} 0 & b & 0 \\ 0 & 0 & 0 \\ x & 0 & b \\ -y & a & 0 \end{pmatrix},$$

and

$$d_2 = \begin{pmatrix} a & 0 & 0 & -b \\ y & x & 0 & 0 \\ 0 & 0 & y & x \end{pmatrix}.$$

Hence

$$d_2h_2 = \begin{pmatrix} yb & 0 & 0 \\ 0 & yb & 0 \\ 0 & xa & yb \end{pmatrix} \quad \text{and} \quad h_2d_2 = \begin{pmatrix} yb & xb & 0 & 0 \\ 0 & 0 & 0 & 0 \\ \hdashline xa & 0 & yb & 0 \\ 0 & xa & 0 & yb \end{pmatrix}.$$

Thus d_2h_2 is congruent, modulo (xa), to $yb\,\mathrm{Id}$. Furthermore, condition (b) of Definition 1.2.2 is the statement that the two bottom rows in the latter matrix are congruent modulo (xa) to $yb\pi_2$. In the context of the diagram in the definition, with $p = 2$, the fact that the lower left (2×2)-matrix is congruent to 0 modulo $f_1 = xa$ is necessary for the map $h_2d_2 : A_1(2) \longrightarrow A_1(2)$ to induce a map

$$A_1(2)/A_1(1) = B_1(2) \longrightarrow A_1(2)/A_1(1) = B_1(2).$$

1.3 What's in This Book?

Extending the case of a single element, if $f_1, \dots, f_c$ is a regular sequence, S is local with residue field k, and both d and h are *minimal* maps, we will show that a higher matrix factorization allows us to describe the *minimal* free resolutions of the HMF module $M := \mathrm{Coker}(d)$ over the rings S and $R := S/(f_1, \dots, f_c)$. Moreover, we will prove that if S is regular, then every sufficiently high syzygy over the complete intersection $R = S/(f_1, \dots, f_c)$ is an HMF module. In the rest of this section we describe our main results more precisely.

We focus on the case when S is a regular local ring and $R = S/(f_1, \dots, f_c)$ is a complete intersection, although most of our results will be proven in greater generality. We will keep the notation of Definition 1.2.2 throughout.

High Syzygies are Higher Matrix Factorization Modules

The next result was the key motivation for our definition of a higher matrix factorization. A more precise version of this result is proved in Corollary 6.4.3.

Theorem 1.3.1 *Let S be a regular local ring with infinite residue field, and let $I \subset S$ be an ideal generated by a regular sequence of length c. Set $R = S/I$, and suppose* *that N is a finitely generated R-module. Let $f_1, \ldots, f_c$ be a generic choice of elements minimally generating I. If M is a sufficiently high syzygy of N over R, then M is the HMF module of a minimal higher matrix factorization (d, h) with respect to $f_1, \ldots, f_c$. Moreover $d \otimes R$ and $h \otimes R$ are the first two differentials in the minimal free resolution of M over R.*

The meaning of "a sufficiently high syzygy" is explained in Sect. 6.1, where we introduce a class of R-modules that we call *pre-stable syzygies* and show that they have the property given in Theorem 1.3.1. Given an R-module N we give in Corollary 6.4.3 a sufficient condition, in terms of $\mathrm{Ext}_R(N, k)$, for the r-th syzygy module of N to be pre-stable. We also explain more about the genericity condition. Over a local Gorenstein ring, we introduce the concept of a stable syzygy in Sect. 6.1 and discuss it in Sect. 7.2.

Minimal R-Free and S-Free Resolutions

Theorem 1.3.1 shows that in order to understand the eventual behavior of minimal free resolutions over the complete intersection R it suffices to construct the minimal free resolutions of HMF modules. This is accomplished by Construction 5.1.1 and Theorem 5.1.2.

The finite minimal free resolution over S of an HMF module is given by Construction 3.1.3 and Theorem 3.1.4. Here is an outline of the codimension 2 case: Let (d, h) be a codimension 2 higher matrix factorization. We first choose splittings

$$A_s(2) = B_s(1) \oplus B_s(2) .$$

Since $d(B_1(1)) \subset B_0(1)$, we can represent the differential d as

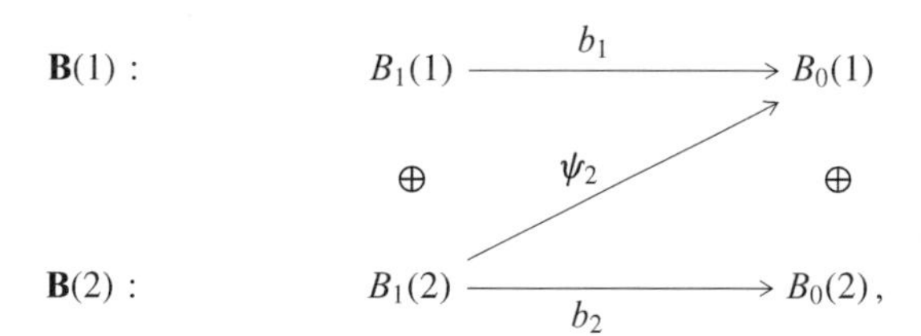

which may be thought of as a map of two-term complexes

$$\psi_2 : \mathbf{B}(2)[-1] \longrightarrow \mathbf{B}(1) .$$

This extends to a map of complexes

$$\mathbf{K}(f_1) \otimes \mathbf{B}(2)[-1] \longrightarrow \mathbf{B}(1) ,$$

as in the following diagram:

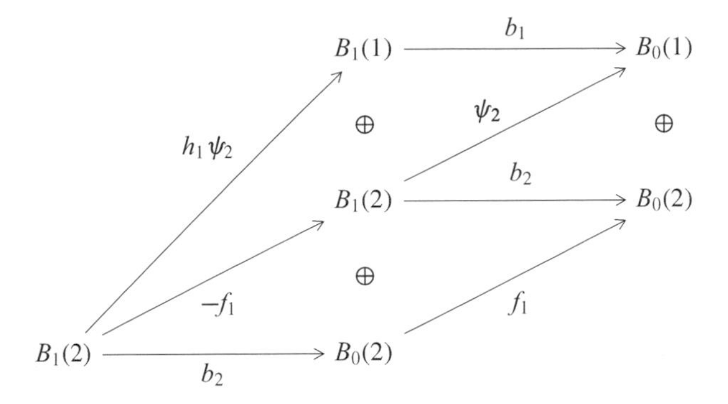

Theorem 3.1.4 asserts that this is the minimal S-free resolution of the HMF module $M = \mathrm{Coker}(S/(f_1,f_2) \otimes d)$.

Strong restrictions on the finite minimal S-free resolution of a high syzygy M over the complete intersection $S/(f_1,\dots,f_c)$ follow from our description: for example, by Corollary 3.2.4 the minimal presentation matrix of M must include $c-1$ columns of the form

$$\begin{pmatrix} f_1 & \cdots & f_{c-1} \\ 0 & \cdots & 0 \\ \vdots & & \vdots \\ 0 & \cdots & 0 \end{pmatrix}$$

for a generic choice of $f_1,\dots,f_c$. For instance, in Example 1.2.4, where $c = 2$, the presentation matrix of M is

$$\begin{pmatrix} a & 0 & 0 & -b & 0 \\ y & x & 0 & 0 & 0 \\ 0 & 0 & y & x & xa \end{pmatrix},$$

and the last column is of the desired type. There are numerical restrictions as well; see Corollary 6.5.3 and the remark following it.

Remark 1.3.2 Every maximal Cohen-Macaulay $S/(f_1)$-module is a pre-stable syzygy, but this is not true in higher codimension—one must go further back in the syzygy chain. This is not surprising, since *every* S-module of finite length is a maximal Cohen-Macaulay module over an artinian complete intersection, and it seems hopeless to characterize the minimal free resolutions of all such modules.

In Corollaries 3.2.6 and 5.2.1 we get formulas for the Betti numbers of an HMF module over S and over R respectively. Furthermore, the graded vector spaces

$\mathrm{Ext}_S(M,k) := \oplus_i \mathrm{Ext}_S(M,k)$ and $\mathrm{Ext}_R(M,k) := \oplus_i \mathrm{Ext}_R(M,k)$ can be expressed as follows:

Corollary 1.3.3 *Suppose that $f_1,\dots,f_c$ is a regular sequence in a regular local ring S with infinite residue field k, so that $R = S/(f_1,\dots,f_c)$ is a local complete intersection. Let M be the HMF module of a minimal higher matrix factorization (d,h) with respect to $f_1,\dots,f_c$. Using notation as in Definition* 1.2.2, *choose splittings $A_s(p) = A_s(p-1) \oplus B_s(p)$ for $s = 0, 1$ and $p = 1,\dots,c$, so that*

$$A_s(p) = \oplus_{1 \le q \le p} B_s(q)\,.$$

Set $B(p) = B_1(p) \oplus B_0(p)$, where we think of $B_s(p)$ as placed in homological degree s. There are decompositions

$$\mathrm{Ext}_S(M,k) \cong \bigoplus_{p=1}^{c} k\langle e_1,\dots,e_{p-1}\rangle \otimes \mathrm{Hom}_S(B(p),k)$$

$$\mathrm{Ext}_R(M,k) \cong \bigoplus_{p=1}^{c} k[\chi_p,\dots,\chi_c] \otimes \mathrm{Hom}_S(B(p),k),$$

as vector spaces, where $k\langle e_1,\dots,e_{p-1}\rangle$ denotes the exterior algebra on variables of degree 1 *and $k[\chi_p,\dots,\chi_c]$ denotes the polynomial ring on variables of degree* 2.

The former formula in Corollary 1.3.3 follows from Remark 3.1.5 and the latter from Corollary 5.1.6. We explain in [28] and Corollary 5.1.6 how the given decompositions reflect certain natural actions of the exterior and symmetric algebras on the graded modules $\mathrm{Ext}_S(M,k)$ and $\mathrm{Ext}_R(M,k)$.

The package CompleteIntersectionResolutions, available in the Macaulay2 system starting with version 1.8, can compute examples of many of the constructions in this book.

Syzygies over Intermediate Quotient Rings

For each $0 \le p \le c$ set

$$R(p) := S/(f_1,\dots,f_p)\,.$$

In the case of a codimension 1 matrix factorization (d,h), one can use the data of the matrix factorization to describe two minimal free resolutions, as explained in (1.2). In the case of a codimension c higher matrix factorization we construct the minimal free resolutions of its HMF module over all $c+1$ rings

$$S = R(0),\ S/(f_1) = R(1),\ \dots,\ S/(f_1,\dots,f_c) = R(c)\,.$$

See Theorem 5.4.4.

By Definition 1.2.2 an HMF module M with respect to the regular sequence $f_1, \ldots, f_c$ determines, for each $p \leq c$, an HMF $R(p)$-module $M(p)$ with respect to $f_1, \ldots, f_p$. In the notation and hypotheses as in Theorem 1.3.1, we have the following properties of the modules $M(p)$: Proposition 7.2.2 shows that

$$M(p-1) = \mathrm{Syz}_2^{R(p-1)}\Big(\mathrm{Syz}_{-2}^{R(p)}\big(M(p)\big)\Big),$$

where $\mathrm{Syz}_i(-)$ and $\mathrm{Syz}_{-i}(-)$ denote syzygy and cosyzygy, respectively. Theorem 7.4.1(2) expresses the modules $M(p)$ as syzygies of the R-module $P := \mathrm{Syz}_{-c-1}^R(M)$ over the intermediate rings $R(p)$ as

$$M(p) = \mathrm{Syz}_{c+1}^{R(p)}(P).$$

Furthermore, Proposition 7.4.2 says that if we replace M by its first syzygy, then all the modules $M(p)$ are replaced by their first syzygies:

$$\Big(\mathrm{Syz}_1^R(M)\Big)(p) = \mathrm{Syz}_1^{R(p)}\Big(M(p)\Big).$$

1.4 Notation and Conventions

Unless otherwise stated, **all rings are assumed commutative and Noetherian, and all modules are assumed finitely generated**.

A map $\phi : A \longrightarrow B$ of S-modules is called *minimal* if S is local and $\phi(A) \subset \mathbf{m}B$, where $\mathbf{m}$ is the maximal ideal of S.

To distinguish a matrix factorization for one element from the general concept, sometimes we will refer to the former as a *codimension* 1 *matrix factorization* or a *hypersurface matrix factorization*.

We will frequently use the following notation about higher matrix factorizations.

Notation 1.4.1 A higher matrix factorization

$$\Big(d : A_1 \longrightarrow A_0,\ h : \oplus_{p=1}^c A_0(p) \longrightarrow A_1\Big)$$

with respect to $f_1, \ldots, f_c$ as in Definition 1.2.2 involves the following data:

- a ring S over which A_0 and A_1 are free modules;
- for $1 \leq p \leq c$, the rings $R(p) := S/(f_1, \ldots, f_p)$, and in particular $R = R(c)$;
- for $s = 0, 1$, the filtrations

$$0 = A_s(0) \subseteq \cdots \subseteq A_s(c) = A_s,$$

preserved by d;

- the induced maps

$$A_0(p) \xrightarrow{h_p} A_1(p) \xrightarrow{d_p} A_0(p);$$

- the quotients $B_s(p) = A_s(p)/A_s(p-1)$ and the projections

$$\pi_p : A_1(p) \longrightarrow B_1(p)\ ;$$

- the two-term complexes induced by d:

$$\mathbf{A}(p):\ A_1(p) \xrightarrow{d_p} A_0(p)$$

$$\mathbf{B}(p):\ B_1(p) \xrightarrow{b_p} B_0(p)$$

- the modules

$$M(p) = \operatorname{Coker}\Big(R(p) \otimes d_p :\ R(p) \otimes A_1(p) \longrightarrow R(p) \otimes A_0(p)\Big),$$

  and in particular, the HMF module $M = M(c)$ of (d, h).

We sometimes write $h = (h_1|\cdots|h_c)$. We say that the higher matrix factorization is *trivial* if $A_1 = A_0 = 0$.

If $1 \leq p \leq c$ then d_p together with the maps h_q for $q \leq p$, is a higher matrix factorization with respect to $f_1, \ldots, f_p$; we write it as $(d_p, h(p))$, where $h(p) = (h_1|\cdots|h_p)$. We call (d_1, h_1) the *codimension* 1 *part* of the higher matrix factorization; (d_1, h_1) is a hypersurface matrix factorization for f_1 over S (it could be trivial). If $q \geq 1$ is the smallest number such that $A(q) \neq 0$ and $R' = S/(f_1, \ldots, f_{q-1})$, then writing $-'$ for $R' \otimes -$, the maps

$$b'_q :\ B_1(q)' \longrightarrow B_0(q)' \ \text{ and } \ h'_q :\ B_0(q)' \longrightarrow B_1(q)'$$

form a hypersurface matrix factorization for the element $f_q \in R'$. We call it the *top non-zero part* of the higher matrix factorization (d, h).

For each $0 \leq p \leq c$ set $R(p) := S/(f_1, \ldots, f_p)$. The HMF module

$$M(p) = \operatorname{Coker}(R(p) \otimes d_p)$$

is an $R(p)$-module.

Next, we make some conventions about complexes which we use throughout.

Conventions on Complexes 1.4.2 We write $\mathbf{U}[-a]$ for the *shifted complex* with

$$\mathbf{U}[-a]_i = \mathbf{U}_{i+a}$$

and differential $(-1)^a d$.

Let $(\mathbf{W}, \partial^W)$ and $(\mathbf{Y}, \partial^Y)$ be complexes. The complex $\mathbf{W} \otimes \mathbf{Y}$ has differential

$$\partial_q^{W \otimes Y} = \sum_{i+j=q} \left((-1)^j \partial_i^W \otimes \mathrm{Id} + \mathrm{Id} \otimes \partial_j^Y\right).$$

If $\varphi : \mathbf{W}[-1] \longrightarrow \mathbf{Y}$ is a map of complexes, so that $-\varphi\partial^W = \partial^Y \varphi$, then the *mapping cone* $\mathbf{Cone}(\varphi)$ is the complex $\mathbf{Cone}(\varphi) = \mathbf{Y} \oplus \mathbf{W}$ with modules

$$\mathbf{Cone}(\varphi)_i = Y_i \oplus W_i$$

and differential

$$\begin{array}{c c} & \begin{array}{cc} Y_i & W_i \end{array} \\ \begin{array}{c} Y_{i-1} \\ W_{i-1} \end{array} & \begin{pmatrix} \partial_i^Y & \varphi_{i-1} \\ 0 & \partial_i^W \end{pmatrix} \end{array}.$$

A map of complexes of free modules $\gamma : \mathbf{W}[a] \longrightarrow \mathbf{Y}$ is homotopic to 0 if there exists a map $\alpha : \mathbf{W}[a+1] \longrightarrow \mathbf{Y}$ such that

$$\gamma = \partial^{\mathbf{Y}}\alpha - \alpha\partial^{\mathbf{W}[a+1]} = \partial^{\mathbf{Y}}\alpha - (-1)^{a+1}\alpha\partial^{\mathbf{W}}.$$

We say that a complex $(\mathbf{U}, d)$ is a *left complex* if $U_j = 0$ for $j < 0$; thus for example the free resolution of a module is a left complex.

If f is an element in a ring S then we write $\mathbf{K}(f)$ for the two-term Koszul complex $f : eS \longrightarrow S$, where we think of e as an exterior variable. If $(\mathbf{W}, \partial)$ is any complex of S-modules we write

$$\mathbf{K}(f) \otimes \mathbf{W} = e\mathbf{W} \oplus \mathbf{W};$$

it is the mapping cone of the map $\mathbf{W} \longrightarrow \mathbf{W}$ that is $(-1)^i f : W_i \longrightarrow W_i$.

Chapter 2
Matrix Factorizations of One Element

Abstract This chapter presents a quick review of the hypersurface case.

2.1 Matrix Factorizations and Resolutions over a Hypersurface

In this section we review the codimension 1 case treated by Eisenbud in [25]. The reader with some experience in the subject can skip to the next section. Recall that if R is a local ring, then a finitely generated R-module M is called *maximal Cohen-Macaulay* if $\operatorname{depth}(M) = \dim(R)$.

Theorem 2.1.1 *Let $0 \neq f \in S$ be a non-zerodivisor in a local ring, and (d, h) be a minimal matrix factorization for f. Let $b = \operatorname{rank}(A_0) = \operatorname{rank}(A_1)$ in the notation of Definition 1.2.1. Set $R = S/(f)$, and let $M = \operatorname{Coker}(d)$ be the matrix factorization R-module.*

(1) *The minimal S-free resolution of M is*

$$0 \longrightarrow S^b \xrightarrow{d} S^b ,$$

and h is a homotopy for f.

(2) *The minimal R-free resolution of M is*

$$\cdots \xrightarrow{R\otimes h} R \otimes A_1 \xrightarrow{R\otimes d} R \otimes A_0 \xrightarrow{R\otimes h} R \otimes A_1 \xrightarrow{R\otimes d} R \otimes A_0$$

It has constant Betti numbers equal to b, and is periodic of period 2.

(3) *Suppose that the ring S is Cohen-Macaulay. Let N be a finitely generated R-module. Then for $p \geq \operatorname{depth}(R) + 1$, the R-module $\operatorname{Syz}_p^R(N)$ is a maximal Cohen-Macaulay module without any free summands.*

(4) *If S is regular, then every maximal Cohen-Macaulay R-module is a direct sum of a free module and a minimal matrix factorization module.*

(5) *If S is regular, and N is a finitely generated R-module, then for $p \geq \operatorname{depth}(R) + 1$, the R-module $\operatorname{Syz}_p^R(N)$ is the module of a minimal matrix factorization, and so*

D. Eisenbud, I. Peeva, *Minimal Free Resolutions over Complete Intersections*, Lecture Notes in Mathematics 2152, DOI 10.1007/978-3-319-26437-0_2

the truncation $\mathbf{F}_{\geq p}$ *of the minimal free resolution* $\mathbf{F}$ *of* N *is described by a matrix factorization.*

Proof First note that $\operatorname{rank}(A_0) = \operatorname{rank}(A_1)$ since after inverting f, both d and h become invertible.

(1) Let $z \in \operatorname{Ker}(d) \subset S^b$. Then

$$fz = hd(z) = h(0) = 0\,.$$

Since f is a non-zerodivisor, it follows that $z = 0$.

(2) We will prove that $\operatorname{Ker}(R \otimes d) = \operatorname{Im}(R \otimes h)$. Let $z \in S^b$ be an element whose image in R^b is in $\operatorname{Ker}(R \otimes d)$. It follows that $d(z) = fu = dh(u)$ for some $u \in S^b$. Since d is a monomorphism, and $d(z-h(u)) = 0$, we conclude that $z = h(u)$. Thus, $\operatorname{Ker}(R \otimes d) = \operatorname{Im}(R \otimes h)$. By symmetry, $\operatorname{Ker}(R \otimes h) = \operatorname{Im}(R \otimes d)$ holds as well.

(3) Let

$$\mathbf{F} : \cdots \longrightarrow F_1 \longrightarrow F_0$$

be the minimal free resolution of N over R. The existence of short exact sequences

$$0 \longrightarrow \operatorname{Syz}_{i+1}(N) \longrightarrow F_i \longrightarrow \operatorname{Syz}_i(N) \longrightarrow 0$$

shows that the depth of the syzygy modules $\operatorname{Syz}_i(N)$ increases strictly with i until it reaches the maximal possible value depth(R), after which it is constant. Thus, for $j \geq \operatorname{depth}(R)$, the R-module $\operatorname{Syz}_j^R(N)$ is maximal Cohen-Macaulay.

It now suffices to show that for $j \geq \operatorname{depth}(R)$ the module $\operatorname{Syz}_{j+1}^R(N)$ has no free summand. Set $U := \operatorname{Syz}_j^R(N)$. Let $g_1, \ldots, g_q$ be a maximal R-regular and U-regular sequence. The module $\bar{U} := U/(g_1, \ldots, g_q)U$ has minimal free resolution $\mathbf{F}_{\geq j} \otimes \bar{R}$ over the artinian ring $\bar{R} = R/(g_1, \ldots, g_q)$. Therefore, the first syzygy of $\bar{U}$ is the reduction modulo $g_1, \ldots, g_q$ of $\operatorname{Syz}_{j+1}^R(N)$. Since $\mathbf{F}$ is a minimal resolution, $\operatorname{Syz}_1^{\bar{R}}(\bar{U})$ is contained in the maximal ideal times $F_j \otimes \bar{R}$; thus it is annihilated by the socle of $\bar{R}$, and cannot contain a free submodule.

(4) Let U be a maximal Cohen-Macaulay R-module. Since $\operatorname{depth}(R) = \operatorname{depth}(S) - 1$, the Auslander-Buchsbaum formula shows that U has projective dimension 1 over S. Thus, its minimal S-free resolution has the form

$$0 \longrightarrow S^a \xrightarrow{\partial} S^e\,.$$

Since U is annihilated by f, we have a homotopy g for f. It follows that (∂, g) is a matrix factorization.

(5) Combine (3) and (4). □

Remark 2.1.2 Note that if we omit the assumption that the matrix factorization (d, h) is minimal, then (1) and (2) in Theorem 2.1.1 still give an S-free and an R-free (possibly non-minimal) resolutions.

Remark 2.1.3 Either one of the two equations in (1.1) suffices to define a matrix factorization provided that the free modules A_0 and A_1 have the same rank. Here is a quick proof: Suppose that $0 \neq f \in S$ is a non-zerodivisor and (d, h) is a pair of maps of finitely generated free modules

$$A_0 \xrightarrow{h} A_1 \xrightarrow{d} A_0$$

of the same rank with $dh = f \cdot \mathrm{Id}$. After inverting f, both d and h become invertible, and $f^{-1}h$ is a right inverse of d. Since inverses are two-sided, $f^{-1}h$ is also a left inverse; that is, $hd = f\mathrm{Id}$ as well.

Chapter 3
Finite Resolutions of HMF Modules

Abstract The main result of this chapter is the construction of the finite minimal free resolution of a higher matrix factorization module in Theorem 3.1.4. We will prove that any such module is the module of a strong matrix factorization in Sect. 5.3.

3.1 The Minimal S-Free Resolution of a Higher Matrix Factorization Module

We will use the notation in 1.4.1 throughout this section. Suppose that M is the HMF module of a higher matrix factorization (d, h) with respect to a regular sequence $f_1, \ldots, f_c$ in a local ring S. Theorem 3.1.4 expresses the minimal S-free resolution of M as an iterated mapping cone of Koszul extensions, which we will now define.

Definition 3.1.1 Let S be a ring. Let $\mathbf{B}$ and $\mathbf{L}$ be S-free left complexes, with $B_i = 0 = L_i$ for $i < 0$, and let $\psi : \mathbf{B}[-1] \longrightarrow \mathbf{L}$ be a map of complexes. Note that ψ is zero on B_0. Let $\mathbf{K} := \mathbf{K}(f_1, \ldots, f_p)$ be the Koszul complex on $f_1, \ldots, f_p \in S$. An $(f_1 \ldots, f_p)$-*Koszul extension* of ψ is a map of complexes

$$\Psi : \mathbf{K} \otimes \mathbf{B}[-1] \longrightarrow \mathbf{L}$$

extending

$$\mathbf{K}_0 \otimes \mathbf{B}[-1] = \mathbf{B}[-1] \xrightarrow{\psi} \mathbf{L}$$

such that the restriction of Ψ to $\mathbf{K} \otimes B_0$ is zero.

The next proposition shows that Koszul extensions exist in the case we will use.

Proposition 3.1.2 *Let $f_1, \ldots, f_p$ be elements of a ring S. Let $\mathbf{L}$ be a free resolution of an S-module N annihilated by $f_1, \ldots, f_p$. Let*

$$\psi : \mathbf{B}[-1] \longrightarrow \mathbf{L}$$

be a map from an S-free left complex $\mathbf{B}$.

D. Eisenbud, I. Peeva, *Minimal Free Resolutions over Complete Intersections*, Lecture Notes in Mathematics 2152, DOI 10.1007/978-3-319-26437-0_3

(1) *There exists an* $(f_1 \ldots, f_p)$*-Koszul extension of* ψ.
(2) *If* S *is local, the elements* f_i *are in the maximal ideal,* $\mathbf{L}$ *is minimal, and the map* ψ *is minimal, then every Koszul extension of* ψ *is minimal.*

Proof Set $\mathbf{K} = \mathbf{K}(f_1 \ldots, f_p)$, and let $\varphi : \mathbf{K} \otimes \mathbf{L} \longrightarrow \mathbf{L}$ be any map extending the identity map $S/(f_1, \ldots f_p) \otimes N \longrightarrow N$. The map φ composed with the tensor product map $\mathrm{Id}_{\mathbf{K}} \otimes \psi$ is a Koszul extension, proving existence.

For the second statement, note that if ψ is minimal, then so is the Koszul extension we have constructed. Since any two extensions of a map from a free complex to a resolution are homotopic, it follows that every Koszul extension is minimal. □

We can now describe how to construct an S-free resolution of an HMF module.

Construction 3.1.3 Let (d, h) be a higher matrix factorization with respect to a regular sequence $f_1, \ldots, f_c$ in a ring S. Using notation as in 1.4.1, we choose splittings

$$A_s(p) = A_s(p-1) \oplus B_s(p)$$

for $s = 0, 1$, so

$$A_s(p) = \oplus_{1 \le q \le p} B_s(q)$$

and denote by ψ_p the component of d_p mapping $B_1(p)$ to $A_0(p-1)$. The construction is by induction on p.

- Let $p = 1$. Set $\mathbf{L}(1) := \mathbf{B}(1)$, a free resolution of $M(1)$ with zero-th term $B_0(1) = A_0(1)$.
- For $p \geq 2$, suppose that $\mathbf{L}(p-1)$ is an S-free resolution of $M(p-1)$ with zero-th term

  $$L_0(p-1) = A_0(p-1)\,.$$

  Let

  $$\psi'_p : \mathbf{B}(p)[-1] \longrightarrow \mathbf{L}(p-1)$$

  be the map of complexes induced by $\psi_p : B_1(p) \longrightarrow A_0(p-1)$, and let

  $$\Psi_p : \mathbf{K}(f_1, \ldots, f_{p-1}) \otimes \mathbf{B}(p)[-1] \longrightarrow \mathbf{L}(p-1)$$

  be an $(f_1, \ldots, f_{p-1})$-Koszul extension. Set

  $$\mathbf{L}(p) = \mathbf{Cone}(\Psi_p).$$

The following theorem implies that $H_0(\mathbf{L}(p)) = M(p)$, so that the construction can be carried through to $\mathbf{L}(c)$. Note that $\mathbf{L}(c)$ has a filtration with successive quotients of the form $\mathbf{K}(f_1, \ldots, f_{p-1}) \otimes \mathbf{B}(p)$.

Theorem 3.1.4 *With notation and hypotheses as in Construction* 3.1.3, *the complex* $\mathbf{L}(p)$ *is an S-free resolution of $M(p)$ for $p = 1, \ldots, c$. Moreover, if S is local and (d, h) is minimal, then the resolution* $\mathbf{L}(p)$ *is minimal.*

Remark 3.1.5 Consider the underlying free module of the Koszul complex $\mathbf{K}(f_1, \ldots, f_{p-1})$ as the exterior algebra on generators e_i corresponding to the f_i. Set $B(p) = B_0(p) \oplus B_1(p)$. As an S-free module $\mathbf{L}(p)$ is

$$\mathbf{L}(p) = \mathbf{L}(p-1) \oplus S\langle e_1, \ldots, e_{p-1}\rangle \otimes_S B(p) .$$

The degree 0 term in $\mathbf{L}$ may be identified with the direct sum of the $B_0(p)$. Note that the only non-zero components of the differential in $\mathbf{L}$ that land in $B_0(p)$ are the map d and the

$$f_i : e_i B_0(p) \longrightarrow B_0(p) \quad \text{for } i < p .$$

The degree 1 term in $\mathbf{L}$ may be identified with the direct sum of the $B_1(p)$ and the $e_i B_0(p)$ for $i < p$. The only non-zero components of the differential in $\mathbf{L}(p)$ that land in $B_1(p)$ are those of the map d and

$$f_i : e_i B_1(p) \longrightarrow B_1(p) \quad \text{for } i < p .$$

However, in $\mathbf{L}$ there could be more such components.

Example 3.1.6 Here is the case of codimension 2. After choosing splittings

$$A_s(2) = B_s(1) \oplus B_s(2) ,$$

a higher matrix factorization (d, h) for a regular sequence $f_1, f_2 \in S$ is a diagram of free S-modules

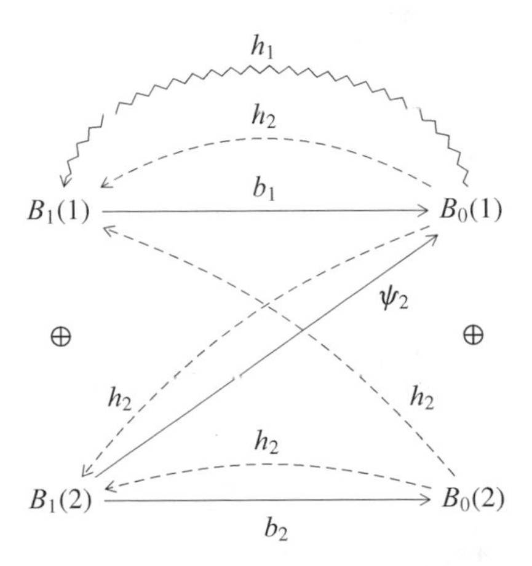

where d has components b_1, b_2, ψ_2, and

$$\begin{aligned} b_1 h_1 &= f_1 \mathrm{Id}_{B_0(1)} \quad \text{on } B_0(1) \\ h_1 b_1 &= f_1 \mathrm{Id}_{B_1(1)} \quad \text{on } B_1(1) \\ dh_2 &\equiv f_2 \mathrm{Id} \mod f_1 (B_0(1) \oplus B_0(2)) \quad \text{on } B_0(1) \oplus B_0(2) \\ \pi_2 h_2 d &\equiv f_2 \pi_2 \mod f_1 B_1(2) \quad \text{on } B_1(1) \oplus B_1(2)\,, \end{aligned} \tag{3.1}$$

where π_2 denotes the projection $B_1(1) \oplus B_1(2) \longrightarrow B_1(2)$. We will see in Sect. 5.3 that the above equations can be simplified by using a strong matrix factorization.

Applying Theorem 3.1.4, we may write the S-free resolution of the HMF module $M = \mathrm{Coker}(S/(f_1, f_2) \otimes d)$ as:

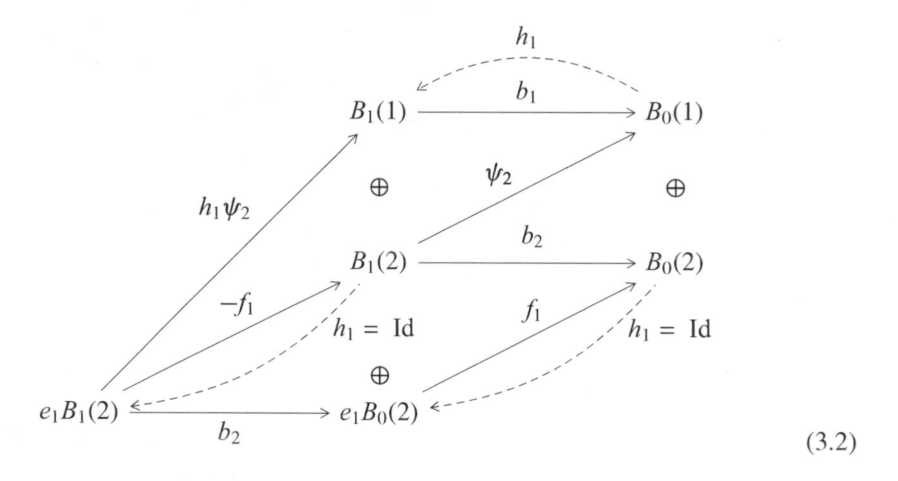

(3.2)

The map h_1 is shown with dashed arrows, and the map h_2 is not shown.

Proof of Theorem 3.1.4 The minimality statement follows at once from the construction and Proposition 3.1.2(2). Thus it suffices to prove the first statement. We will do this by induction on p.

Let $p = 1$. Note that $d_1 = b_1$. The equations in the definition of a higher matrix factorization imply in particular that $h_1 b_1 = b_1 h_1 = f_1 \mathrm{Id}$, so b_1 is a monomorphism. Note that $\mathrm{Coker}(d_1)$ is annihilated by f_1. Thus $\mathbf{L}(1) = \mathbf{B}(1)$ is an S-free resolution of

$$M(1) = \mathrm{Coker}\Big(R(1) \otimes d_1\Big) = \mathrm{Coker}(d_1).$$

For $p \geq 2$, by the induction hypothesis

$$\mathbf{L}(p-1): \quad \cdots \longrightarrow L_1(p-1) \longrightarrow L_0(p-1)$$

is a free resolution of $M(p-1)$. Since $L_0(p-1) = A_0(p-1)$, the map ψ_p defines a morphism of complexes

$$\psi'_p : \mathbf{B}(p)[-1] \longrightarrow \mathbf{L}(p-1)$$

and thus a mapping cone

$$\begin{array}{ccccccc} \cdots & \longrightarrow & L_2(p-1) & \longrightarrow & L_1(p-1) & \longrightarrow & L_0(p-1) \\ & & & & & \overset{\psi_p}{\nearrow} & \\ & & & & B_1(p) & \xrightarrow[b_p]{} & B_0(p)\,. \end{array}$$

To simplify the notation, denote by $\mathbf{K}$ the Koszul complex $\mathbf{K}(f_1,\ldots,f_{p-1})$ of $f_1,\ldots,f_{p-1}$, and write

$$\kappa_i : \wedge^i S^{p-1} \longrightarrow \wedge^{i-1} S^{p-1}$$

for its differential. Also, set $B_s := B_s(p)$ and $\mathbf{B} :\ B_1 \xrightarrow{b_p} B_0$.

Since $M(p-1)$ is annihilated by $(f_1,\ldots,f_{p-1})$, Proposition 3.1.2 shows that there exists a Koszul extension

$$\Psi_p : \mathbf{K} \otimes \mathbf{B}[-1] \longrightarrow \mathbf{L}(p-1)$$

of ψ'_p. Let $(\mathbf{L}(p), \epsilon)$ be the mapping cone of Ψ_p, and note that the zero-th term of $\mathbf{L}(p)$ is

$$L_0 = L_0(p-1) \oplus B_0 = A_0(p)\,.$$

We will show that $\mathbf{L}(p)$ is a resolution of $M(p)$.

We first show that

$$H_0(\mathbf{L}(p)) = \mathrm{Coker}(\epsilon_1) = M(p)\,.$$

If we drop the columns corresponding to B_1 from a matrix for ϵ_1 we get a presentation of

$$M(p-1) \oplus \Big(R(p-1) \otimes B_0\Big)\,,$$

so $\mathrm{Coker}(\epsilon_1)$ is annihilated by $(f_1,\ldots,f_{p-1})$. Moreover, the map

$$h_p : A_0(p) \longrightarrow A_1(p) \subset L_1(p)$$

defines a homotopy for multiplication by f_p modulo $(f_1, \ldots, f_{p-1})$, and so $\mathrm{Coker}(\epsilon_1)$ is annihilated by f_p as well. Thus

$$\mathrm{Coker}(\epsilon_1) = \mathrm{Coker}(R(p) \otimes \epsilon_1) = M(p)$$

as required.

Since $\mathbf{L}(p)$ is the mapping cone of Ψ_p, we have a long exact sequence in homology of the form

$$\cdots \to H_i(\mathbf{L}(p-1)) \longrightarrow H_i(\mathbf{L}(p)) \longrightarrow H_i(\mathbf{K} \otimes \mathbf{B}) \xrightarrow{\Psi_{p*}} H_{i-1}(\mathbf{L}(p-1)) \to \cdots .$$

The complex $\mathbf{K} \otimes \mathbf{B}$ may be thought of as the mapping cone of the map

$$(-1)^i b_p \otimes \mathrm{Id} : \; K_i \otimes B_1[-1] \longrightarrow K_i \otimes B_0 \,,$$

so there is also a long exact sequence

$$\cdots \to H_i(\mathbf{K} \otimes B_1) \longrightarrow H_i(\mathbf{K} \otimes B_0) \longrightarrow H_i(\mathbf{K} \otimes \mathbf{B}) \longrightarrow H_{i-1}(\mathbf{K} \otimes B_1) \to \cdots .$$

Since $\mathbf{K} \otimes B_s$ is a resolution of $R(p-1) \otimes B_s$ we see that $H_i(\mathbf{K} \otimes \mathbf{B}) = 0$ for $i > 1$, which implies that $H_i(\mathbf{L}(p)) = 0$ for $i > 1$.

To show that $H_1(\mathbf{L}(p))$ also vanishes, we will show that the map

$$\Psi_{p*} : \; H_1(\mathbf{K} \otimes \mathbf{B}) \longrightarrow H_0(\mathbf{L}(p-1)) = M(p-1)$$

is a monomorphism. From the long exact sequence for $\mathbf{K} \otimes B$ above we get the four-term exact sequence

$$\begin{aligned} 0 \longrightarrow H_1(\mathbf{K} \otimes \mathbf{B}) \longrightarrow R(p-1) \otimes B_1 &\xrightarrow{R(p-1)\otimes b_p} R(p-1) \otimes B_0 \\ &\longrightarrow H_0(\mathbf{K} \otimes \mathbf{B}) \longrightarrow 0. \end{aligned}$$

Thus

$$H_1(\mathbf{K} \otimes \mathbf{B}) = \mathrm{Ker}\Big(R(p-1) \otimes b_p\Big) .$$

By construction the map

$$\Psi_{p*} : \mathrm{Ker}\Big(R(p-1) \otimes b_p\Big) \longrightarrow H_0(\mathbf{L}(p-1)) = \mathrm{Coker}\Big(R(p-1) \otimes d_{p-1}\Big)$$

is induced by

$$\psi_p : R(p-1) \otimes B_1(p) \longrightarrow R(p-1) \otimes A_0(p-1) .$$

The proof is completed by Lemma 3.1.7, which we will use again in Sect. 5.1. □

Lemma 3.1.7 *With notation and hypotheses as in Construction* 3.1.3, ψ_p *induces a monomorphism from* $\operatorname{Ker}\Big(R(p-1)\otimes b_p\Big)$ *to* $\operatorname{Coker}\Big(R(p-1)\otimes d_{p-1}\Big)$.

Proof To simplify notation we write $\overline{-}$ for $R(p-1)\otimes -$. Consider the diagram:

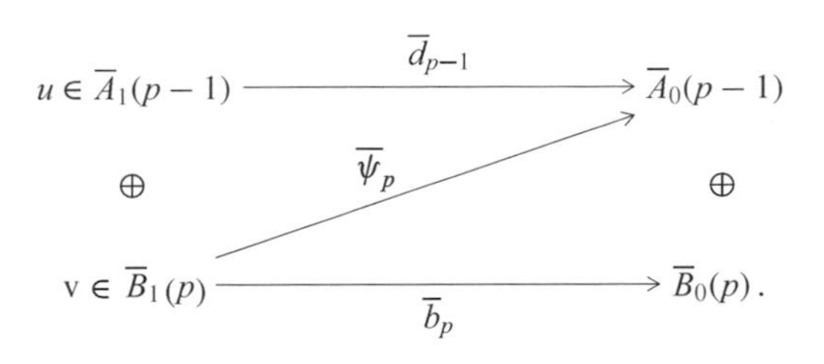

We must show that if $v\in\operatorname{Ker}(\overline{b}_p)$ and $\overline{\psi}_p(v)=\overline{d}_{p-1}(u)$ for some $u\in\overline{A}_1(p-1)$, then $v=0$.

Write $\overline{\pi}_p$ for the projection

$$\overline{\pi}_p:\ \overline{A}_1(p)=\overline{A}_1(p-1)\oplus\overline{B}_1(p)\ \longrightarrow\ \overline{B}_1(p)\,,$$

and note that $\overline{d}_p$ is the sum of the three maps in the diagram above. Our equations say that $d_p(-u,v)=0$. By condition (b) in Definition 1.2.2,

$$f_p v=f_p\overline{\pi}_p(-u,v)=\overline{\pi}_p\overline{h}_p\overline{d}_p(-u,v)=0.$$

Since f_p is a non-zerodivisor in $R(p-1)$, it follows that $v=0$. □

3.2 Consequences

Throughout this section we keep the notation and hypotheses of Construction 3.1.3. Recall that

$$M(p):=\operatorname{Coker}(S/(f_1,\ldots,f_p)\otimes d_p)\,.$$

Theorem 3.1.4 immediately yields:

Corollary 3.2.1 *The S-free resolution* $\mathbf{L}=\mathbf{L}(c)$ *of* M *has a filtration by the S-free resolutions* $\mathbf{L}(p)$ *of the modules* $M(p)$*, whose successive quotients are the complexes*

$$\mathbf{K}(f_1,\ldots,f_{p-1})\otimes_S\mathbf{B}(p).$$

□

Corollary 3.2.2 *If $M(p) \neq 0$, then its projective dimension over S is p. If S is a local Cohen-Macaulay ring, then $M(p)$ is a maximal Cohen-Macaulay $R(p)$-module.*

Proof The resolution $\mathbf{L}(p)$ has length p, and no module annihilated by a regular sequence of length p can have projective dimension $< p$. The Cohen-Macaulay statement follows from this and the Auslander-Buchsbaum formula. □

Corollary 3.2.3 *The element f_{p+1} is a non-zerodivisor on $M(p)$.*

Proof If f_{p+1} is a zerodivisor on $M(p)$ then, since $f_1, \dots, f_p$ annihilate $M(p)$,

$$\operatorname{Tor}^S_{p+1}\Big(S/(f_1, \dots, f_{p+1}), M(p)\Big) = H_{p+1}\Big(\mathbf{K}(f_1, \dots, f_{p+1}) \otimes M(p)\Big) \neq 0.$$

However, the length of the complex $\mathbf{L}(p)$ is only p, so

$$\operatorname{Tor}^S_{p+1}\Big(S/(f_1, \dots, f_{p+1}\Big), M(p)) = 0\,,$$

a contradiction. □

The following result shows that HMF modules are quite special. Looking ahead to Corollary 6.4.3, we see that it can be applied to *any* S module that is a sufficiently high syzygy over R.

Corollary 3.2.4 *Suppose that S is local and that the higher matrix factorization (d, h) is minimal, and let $n = \operatorname{rank} A_0$, the rank of the target of d. In a suitable basis, the minimal presentation matrix of the HMF module M consists of the matrix d concatenated with an $\big(n \times \sum_p (p-1)\operatorname{rank} B_0(p)\big)$-matrix that is the direct sum of matrices of the form*

$$\big(f_1 \dots f_{p-1}\big) \otimes \mathrm{Id}_{B_0(p)} = \begin{pmatrix} f_1 \cdots f_{p-1} & 0 \cdots & 0 & \cdots 0 \cdots & 0 \\ 0 \cdots \quad 0 & f_1 \cdots & f_{p-1} & \cdots 0 \cdots & 0 \\ 0 \cdots \quad 0 & 0 \cdots & 0 & \cdots 0 \cdots & 0 \\ \vdots \quad \vdots \quad \vdots & \vdots \quad \vdots & \vdots & \vdots \quad \vdots \quad \vdots & \vdots \\ 0 \cdots \quad 0 & 0 \cdots & 0 & \cdots f_1 \cdots & f_{p-1} \end{pmatrix}$$

Similar results hold for all differentials in the minimal free resolution $\mathbf{L}$ of the module M.

Proof In the notation of Construction 3.1.3, the given direct sum is the part of the map $\mathbf{L}_1(c) \longrightarrow \mathbf{L}_0(c)$ that corresponds to

$$\oplus_p \Big(\mathbf{K}(f_1, \dots, f_{p-1})\Big)_1 \otimes B_0(p) \longrightarrow \oplus_p B_0(p).$$

□

Corollary 3.2.5 *If S is local and the higher matrix factorization is minimal, then $M(p)$ has no $R(p)$-free summands.*

Proof If $M(p)$ had an $R(p)$-free summand, then with respect to suitable bases the minimal presentation matrix $R(p) \otimes d_p$ of $M(p)$ would have a row of zeros. Thus a matrix representing $R(p-1) \otimes d_p$ would have a row of elements divisible by f_p. Composing with h_p we see that a matrix representing $R(p-1) \otimes d_p h_p$ would have a row of elements in $\mathbf{m} f_p$. However

$$R(p-1) \otimes \left(d_p h_c\right) = f_p \mathrm{Id}\,,$$

a contradiction. □

Recall that if S is a local ring with residue field k, then the *Betti numbers* of a module N over S are

$$\beta_i^S(N) = \dim_k(\mathrm{Tor}_i^S(N,k)) = \dim_k(\mathrm{Ext}_S^i(N,k))\,.$$

They are often studied via the *Poincaré series*

$$\mathcal{P}_N^S(x) = \sum_{i \geq 0} \beta_i^S(N)\, x^i\,.$$

Theorem 3.1.4 allows us to express the Betti numbers and Poincaré series of an HMF module in terms of the ranks of the modules $B_s(p)$:

Corollary 3.2.6 *If S is local and the higher matrix factorization (d, h) is minimal, then*

$$\mathcal{P}_M^S(x) = \sum_{1 \leq p \leq c} (1+x)^{p-1}\left(x \operatorname{rank}(B_1(p)) + \operatorname{rank}(B_0(p))\right).$$

It is worthwhile to ask whether there are interesting restrictions on the ranks of the $B_s(p)$. Here is a first result in this direction:

Corollary 3.2.7 *If S is local and the higher matrix factorization (d, h) is minimal, then*

$$\begin{aligned} \operatorname{rank} B_1(p) \;\geq\; & \operatorname{rank} B_0(p) \\ & + \max\left\{ \frac{\text{length } M(p-1)_Q}{\text{length } R(p-1)_Q} \;\middle|\; Q \text{ is a minimal prime of } R(p-1) \right\} \end{aligned}$$

for each $p = 1, \ldots, c$. Thus

$$\operatorname{rank} B_1(p) \geq \operatorname{rank} B_0(p)$$

for every p. Furthermore, $\operatorname{rank} B_1(p) = \operatorname{rank} B_0(p)$ *if and only if* $M(p-1) = 0$*; in this case,* $B_s(q) = 0$ *for all* $q < p,\ s = 0, 1$.

Proof The module $M(p)$ is annihilated by f_p. Since f_p is a non-zerodivisor in $R(p-1)$, it follows that the localization $M(p)_Q$ is 0 for any minimal prime Q of $R(p-1)$. Note that $M(p)$ is obtained from $M(p-1) \oplus \big(R(p-1) \otimes B_0(p)\big)$ by factoring out the image of $R(p-1) \otimes B_1(p)$ under the map with components $R(p-1) \otimes \psi_p$ and $R(p-1) \otimes b_p$. After localizing at Q we get the desired inequality

$$\big(\text{length } R(p-1)_Q\big) \cdot \operatorname{rank} B_1(p)_Q \geq \big(\text{length } R(p-1)_Q\big) \cdot \operatorname{rank} B_0(p)_Q + \text{length } M(p-1)_Q$$

since the length of a free $R(p-1)_Q$-module is the rank times the length of $R(p-1)_Q$.

If $M(p-1) = 0$, then by the minimality of the matrix factorization for $M(p-1)$ we have $A_s(p-1) = 0$, and thus $B_s(q) = 0$ for $s = 1, 2$ and $q < p$. In this case the equations in the definition of a matrix factorization imply that $\big(R(p-1) \otimes d_p, R(p-1) \otimes h_p\big)$ is a codimension 1 matrix factorization of the image of f_p in $R(p-1)$, whence

$$\operatorname{rank} B_1(p) = \operatorname{rank} B_0(p)\,.$$

Conversely, suppose $M(p-1) \neq 0$. Suppose $M(p-1)_Q = 0$ for every minimal prime $Q \supseteq (f_1, \dots, f_{p-1})$ of S. Then the height of $\operatorname{Ann}\big(M(p-1)\big)$ is greater than $p-1$. The projective dimension of $M(p-1)_Q$ over S_Q is less or equal to $p-1$, so it is strictly less than $\dim(S_Q)$. Thus the minimal S_Q-free resolution of $M(p-1)_Q$ is a complex of length $< \dim(S_Q)$ and its homology $M(p-1)_Q$ has finite length. This is a contradiction by the New Intersection Theorem, cf. [49]. Thus, there exists a minimal prime Q' of $R(p-1)$ for which $M(p-1)_{Q'} \neq 0$. $\square$

Example 3.2.8 Let $S = k[x, y, z]$ and let f_1, f_2 be the regular sequence xz, y^2. We give an example of a higher matrix factorization with respect to f_1, f_2 such that $B_1(2) \neq 0$, but $B_0(2) = 0$. If

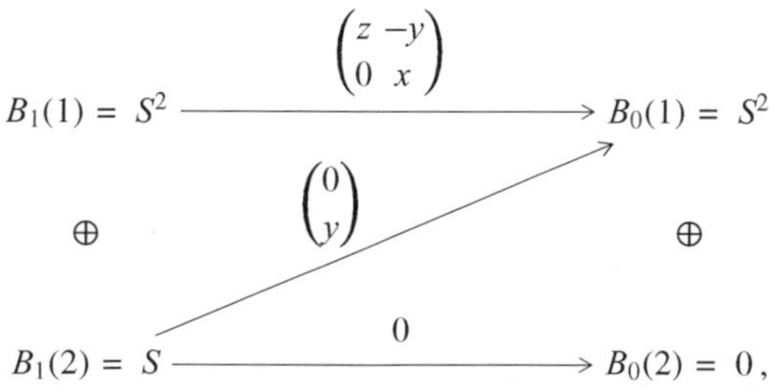

$$\begin{array}{ccc} B_1(1) = S^2 & \xrightarrow{\begin{pmatrix} z & -y \\ 0 & x \end{pmatrix}} & B_0(1) = S^2 \\ \oplus & \nearrow_{\begin{pmatrix} 0 \\ y \end{pmatrix}} & \oplus \\ B_1(2) = S & \xrightarrow{\quad 0 \quad} & B_0(2) = 0\,, \end{array}$$

and

$$h_1 = \begin{pmatrix} x & y \\ 0 & z \end{pmatrix}$$

$$h_2 = \begin{pmatrix} 0 & 0 \\ -y & 0 \\ x & y \end{pmatrix},$$

then (d, h) is a higher matrix factorization.

In the case of higher matrix factorizations that come from high syzygies (stable matrix factorizations) Corollary 3.2.7 can be strengthened further: $B_0(p) = 0$ implies $B_1(p) = 0$ as well; see Corollary 6.5.1. This is not the case in general, as the above example shows.

3.3 Building a Koszul Extension

It is possible to give rather explicit formulas for the maps in a Koszul extension, and thus to express the differentials in the resolution **L** of Theorem 3.1.4 inductively in terms of d and h applying Proposition 3.3.1; we use this to study the structure of Tor and Ext ([28] and a project in progress). We will not use Proposition 3.3.1 in this book, so the section may be skipped without loss of continuity.

Proposition 3.3.1 *Suppose that $\psi : \mathbf{W}[-1] \longrightarrow \mathbf{Y}$ is a map of free left complexes over S, and ν is a homotopy for multiplication by $f \in S$ on $\mathbf{Y}$.*

(1) *The map*

$$\Psi_i = ((-1)^{i-1}\nu_{i-2}\psi_{i-2}, \psi_{i-1}) : eW_{i-1} \oplus W_i \longrightarrow Y_{i-1}$$

defines an f-Koszul extension of ψ. Thus, we have the following diagram of the complex $\mathbf{Cone}(\Psi)$*, where we have denoted ∂ and δ the differentials in $\mathbf{Y}$ and $\mathbf{W}$, respectively:*

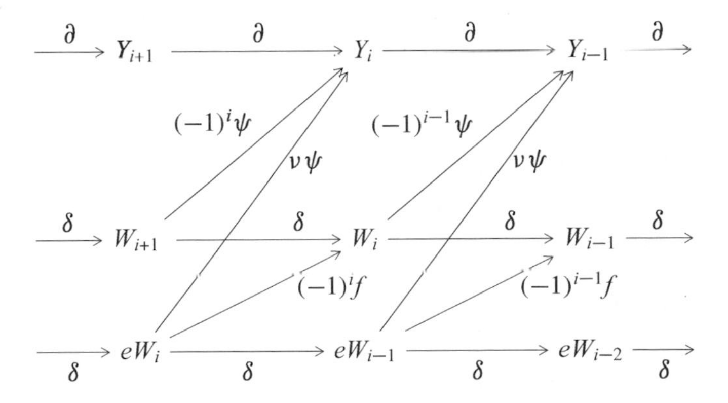

(2) *If* $-\tau$ *is a homotopy for* $\nu^2 \sim 0$ *on* $\mathbf{Y}$, *then*

$$\xi := (\tau\psi : eW_{i-1} \longrightarrow Y_{i+1},\ (-1)^i : W_i \longrightarrow eW_i,\ \nu : Y_i \longrightarrow Y_{i+1})$$

is a homotopy for multiplication by f *on the mapping cone* $\mathbf{Cone}(\Psi)$. *We have the following diagram of that homotopy:*

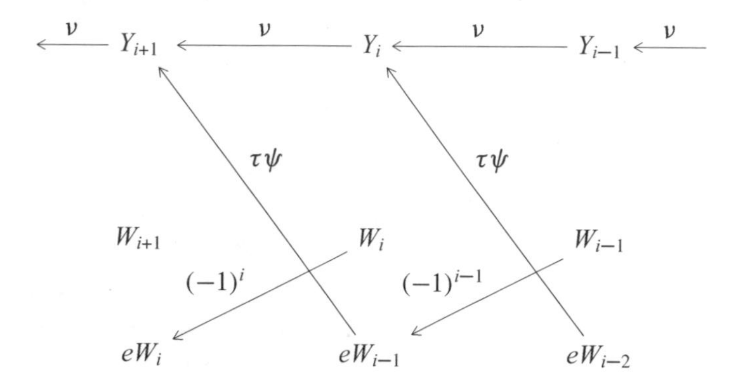

The restriction of the homotopy ξ *on* $\mathbf{Y}$ *is the given homotopy* ν.

Proof (1) and (2) are proven by "chasing" the given diagrams, using the formulas:

$$\partial\psi = \psi\delta \text{ (since } \psi \text{ is a map of complexes)}$$
$$\nu^2 + \partial\tau + \tau\partial = 0 \text{ (by the definition of } \tau)$$
$$\nu^2\psi + \partial\tau\psi + \tau\psi\delta = 0 \text{ (by the above two formulas)}.$$

□

Definition 3.3.2 Under the assumptions and in the notation in the above lemma, we call $\Psi : \mathbf{K}(f) \otimes \mathbf{W}[-1] \longrightarrow \mathbf{Y}$ the *distinguished extension* of ψ and denote it $\mathbf{de}(\psi, \nu)$; we call $\mathbf{Cone}(\Psi)$ the *distinguished mapping cone*.

3.4 Higher Homotopies

In this section we recall the concept of higher homotopies. The version for a single element is due to Shamash [55]; Eisenbud [24] treats the more general case of a collection of elements.

Definition 3.4.1 Let $f_1, \ldots, f_c \in S$, and $\mathbf{G}$ be a complex of free S-modules. We denote $\mathbf{a} = (a_1, \ldots, a_c)$, where each $a_i \geq 0$ is an integer, and set $|\mathbf{a}| = \sum_i a_i$. A *system of higher homotopies* σ for $f_1, \ldots, f_c$ on $\mathbf{G}$ is a collection of maps

$$\sigma_{\mathbf{a}} : \mathbf{G} \longrightarrow \mathbf{G}[-2|\mathbf{a}| + 1]$$

of the underlying modules such that the following three conditions are satisfied:

(1) $\sigma_{\mathbf{0}}$ is the differential on $\mathbf{G}$.
(2) For each $1 \le i \le c$, the map $\sigma_{\mathbf{0}}\sigma_{\mathbf{e}_i} + \sigma_{\mathbf{e}_i}\sigma_{\mathbf{0}}$ is multiplication by f_i on $\mathbf{G}$, where $\mathbf{e}_i$ is the i-th standard vector.
(3) If $\mathbf{a}$ is a multi-index with $|\mathbf{a}| \ge 2$, then

$$\sum_{\mathbf{b}+\mathbf{s}=\mathbf{a}} \sigma_{\mathbf{b}}\sigma_{\mathbf{s}} = 0\,.$$

A system of higher homotopies σ for one element $f \in S$ on $\mathbf{G}$ consists of maps

$$\sigma_j : \mathbf{G} \longrightarrow \mathbf{G}[-2j+1]$$

for $j = 0, 1, \ldots$, and will be denoted $\{\sigma_j\}$.

Proposition 3.4.2 ([24, 55]) *If $\mathbf{G}$ is a free resolution of an S-module annihilated by elements $f_1, \ldots, f_c \in S$, then there exists a system of higher homotopies on $\mathbf{G}$ for $f_1, \ldots, f_c$.*

For the reader's convenience we present a short proof following the idea in [55]:

Proof It is well-known that homotopies $\sigma_{\mathbf{e}_i}$ satisfying (2) in Definition 3.4.1 exist. The equation in item (3) in Definition 3.4.1 can be written as:

$$d\sigma_{\mathbf{a}} = -\sum_{\substack{\mathbf{b}+\mathbf{s}=\mathbf{a}\\ \mathbf{b}\neq\mathbf{0}}} \sigma_{\mathbf{b}}\sigma_{\mathbf{s}}\,.$$

As $\mathbf{G}$ is a free resolution, in order to show by induction on $\mathbf{a}$ and on the homological degree that the desired $\sigma_{\mathbf{a}}$ exists, it suffices to show that the right-hand side is annihilated by d. Indeed:

$$\begin{aligned}
-\sum_{\substack{\mathbf{b}+\mathbf{s}=\mathbf{a}\\ \mathbf{b}\neq\mathbf{0}}} (d\sigma_{\mathbf{b}})\sigma_{\mathbf{s}} &= \sum_{\substack{\mathbf{b}+\mathbf{s}=\mathbf{a}\\ \mathbf{b}\neq\mathbf{0}}} \sum_{\substack{\mathbf{m}+\mathbf{r}=\mathbf{b}\\ \mathbf{r}\neq\mathbf{0}}} \sigma_{\mathbf{r}}\sigma_{\mathbf{m}}\sigma_{\mathbf{s}} - \sum_{\{i:\, \mathbf{e}_i<\mathbf{a}\}} f_i\sigma_{\mathbf{a}-\mathbf{e}_i}\\
&= \sum_{\substack{\mathbf{m}+\mathbf{r}+\mathbf{s}=\mathbf{a}\\ \mathbf{r}\neq\mathbf{0}}} \sigma_{\mathbf{r}}\sigma_{\mathbf{m}}\sigma_{\mathbf{s}} - \sum_{\{i:\, \mathbf{e}_i<\mathbf{a}\}} f_i\sigma_{\mathbf{a}-\mathbf{e}_i}\\
&= -\sum_{\{i:\, \mathbf{e}_i<\mathbf{a}\}} f_i\sigma_{\mathbf{a}-\mathbf{e}_i} + \sum_{\mathbf{r}\neq\mathbf{0}} \sigma_{\mathbf{r}}\Bigg(\sum_{\mathbf{m}+\mathbf{s}=\mathbf{a}-\mathbf{r}} \sigma_{\mathbf{m}}\sigma_{\mathbf{s}}\Bigg)\\
&= \sum_{\substack{\mathbf{r}\neq\mathbf{0}\\ \mathbf{r}\neq\mathbf{a}-\mathbf{e}_i}} \sigma_{\mathbf{r}}\Bigg(\sum_{\mathbf{m}+\mathbf{s}=\mathbf{a}-\mathbf{r}} \sigma_{\mathbf{m}}\sigma_{\mathbf{s}}\Bigg) + \sum_{\{i:\, \mathbf{e}_i<\mathbf{a}\}} \sigma_{\mathbf{a}-\mathbf{e}_i}(\sigma_{\mathbf{e}_i}\sigma_{\mathbf{0}} + \sigma_{\mathbf{0}}\sigma_{\mathbf{e}_i} - f_i) = 0\,,
\end{aligned}$$

where the first and the last equalities hold by induction hypothesis. □

Chapter 4
CI Operators

Abstract In this chapter we discuss basic properties of CI operators.

4.1 CI Operators

In this section, we review material from [25].

Construction 4.1.1 ([25]) Suppose that $f_1, \ldots, f_c \in S$ and $(\mathbf{V}, \partial)$ is a complex of free modules over $R = S/(f_1, \ldots, f_c)$. Let $\tilde{\partial}$ be a lifting of ∂ to S, that is, a sequence of S-free modules $\widetilde{V}_i$ and maps

$$\tilde{\partial}_{i+1} : \widetilde{V}_{i+1} \longrightarrow \widetilde{V}_i\,,$$

such that $\partial = R \otimes \tilde{\partial}$. Since $\partial^2 = 0$ we can choose maps

$$\widetilde{t_j} : \widetilde{V}_{i+1} \longrightarrow \widetilde{V}_{i-1}\,,$$

where $1 \leq j \leq c$, such that

$$\tilde{\partial}^2 = \sum_{j=1}^{c} f_j \tilde{t}_j.$$

The *CI operators* (sometimes called Eisenbud operators) associated to the sequence $f_1, \ldots, f_c$ are

$$t_j := R \otimes \widetilde{t_j}\,.$$

In the case $c = 1$, we have $\tilde{\partial}^2 = f_1 \tilde{t}_1$ and we sometimes write $\tilde{t}_1 = \frac{1}{f_1}\tilde{\partial}^2$ and call it the *lifted CI operator*.

D. Eisenbud, I. Peeva, *Minimal Free Resolutions over Complete Intersections*, Lecture Notes in Mathematics 2152, DOI 10.1007/978-3-319-26437-0_4

The next result gives some basic properties of the CI operators proven in [25]. For the reader's convenience we provide its proof.

Proposition 4.1.2 ([25]) *Suppose that $f_1, \dots, f_c \in S$ is a regular sequence, $\alpha : (\mathbf{V}, \partial) \longrightarrow (\mathbf{W}, \delta)$ is a map of complexes of free modules over $R = S/(f_1, \dots, f_c)$, and $\tilde{\partial}$ and $\tilde{\delta}$ are liftings to S of ∂ and δ, respectively.*

(1) *The CI operators on $\mathbf{V}$ are maps of complexes $\mathbf{V}[-2] \longrightarrow \mathbf{V}$.*
(2) *The CI operators on $\mathbf{V}$ are uniquely determined by the choice of $\tilde{\partial}$.*
(3) *The CI operators on $\mathbf{V}$ are, up to homotopy, independent of the choice of the lifting $\tilde{\partial}$.*
(4) *Up to homotopy, the CI operators commute with α.*

Proof The proofs of these statements all follow the same pattern.

(1) To say that t_j is a map of complexes means that it commutes with ∂. The equality

$$\sum_{j=1}^{c} f_j \tilde{t}_j \tilde{\partial} = \widetilde{\partial}^3 = \tilde{\partial} \sum_{j=1}^{c} f_j \tilde{t}_j,$$

implies

$$\sum_{j=1}^{c} f_j \Big(\tilde{t}_j \tilde{\partial} - \tilde{\partial} \tilde{t}_j \Big) = 0 .$$

Since $f_1, \dots, f_c$ is a regular sequence, this implies $\tilde{t}_j \tilde{\partial} \equiv \tilde{\partial} \tilde{t}_j \mod (f_1, \dots, f_c)$. Thus, $t_j \partial = \partial t_j$ as required.

(2) If $\tilde{\partial}^2 = \sum_{j=1}^{c} f_j \tilde{t}'_j$ is another expression, then

$$\sum_{j=1}^{c} f_j (\tilde{t}'_j - \tilde{t}_j) = 0 .$$

Hence, $\tilde{t}'_j \equiv \tilde{t}_j \mod (f_1, \dots, f_c)$, which implies $t'_j = t_j$.

(4) Denote t_j^V and t_j^W the CI operators on $\mathbf{V}$ and $\mathbf{W}$, respectively. To show that the CI operators commute with α up to homotopy, note that since α is a map of complexes we may write

$$\tilde{\delta} \tilde{\alpha} - \tilde{\alpha} \tilde{\partial} = \sum_{j=1}^{c} \tilde{h}_j f_j$$

for some maps $\tilde{h}_j$, whence

$$\begin{aligned}\sum_{j=1}^{c} f_j \tilde{t}_j^W \tilde{\alpha} &= \tilde{\delta}^2 \tilde{\alpha} \\ &= \tilde{\delta}\tilde{\alpha}\tilde{\partial} + \sum_{j=1}^{c} \tilde{\delta}\tilde{h}_j f_j \\ &= \tilde{\alpha}\tilde{\partial}^2 + \sum_{j=1}^{c} \tilde{h}_j f_j \tilde{\partial} + \sum_{j=1}^{c} \tilde{\delta}\tilde{h}_j f_j \\ &= \sum_{j=1}^{c} \left(\tilde{\alpha} f_j \tilde{t}_j^V + \tilde{\delta}\tilde{h}_j f_j + \tilde{h}_j \tilde{\partial} f_j \right).\end{aligned}$$

Thus,

$$\sum_{j=1}^{c} f_j \left(\tilde{t}_j^W \tilde{\alpha} - (\tilde{\alpha}\tilde{t}_j^V + \tilde{\delta}\tilde{h}_j + \tilde{h}_j \tilde{\partial}) \right) = 0 .$$

As before, it follows that $t_j^W \alpha - \alpha t_j^V = \delta h_j + h_j \partial$.

(3) Apply part (4) to the case where $\mathbf{V} = \mathbf{W}$ and α is the identity. □

Note that we have never made any hypothesis that $\mathbf{V}$ is minimal. There is a family of (often) non-minimal complexes for which the action of the CI operators is easy to describe; these are the complexes coming from the Shamash construction, which we now recall. Shamash [55] actually treats only the version with $c = 1$; while Eisenbud [24] treats the more general case.

Construction 4.1.3 (see [25, Sect. 7]) Given elements $f_1, \dots, f_c$ in a ring S, and a free left complex $\mathbf{G}$ over S with a system σ of higher homotopies, we will define a complex $\mathrm{Sh}(\mathbf{G}, \sigma)$ over $R := S/(f_1, \dots, f_c)$.

Write $S\{y_1, \dots, y_c\}$ for the divided power algebra over S on variables $y_1, \dots, y_c$ of degree 2; thus,

$$S\{y_1, \dots, y_c\} \cong \mathrm{Hom}_{\text{graded } S\text{-modules}}(S[t_1, \dots, t_c], S) = \oplus\, S y_1^{(i_1)} \cdots y_c^{(i_c)}$$

where the "divided monomials" $y_1^{(i_1)} \cdots y_c^{(i_c)}$ form the dual basis to the monomial basis of the polynomial ring $S[t_1, \dots, t_c]$. The divided power algebra $S\{y_1, \dots, y_c\}$ is a graded module over $S[t_1, \dots, t_c]$ with action

$$t_j y_j^{(i)} = y_j^{(i-1)} ,$$

(see [26, Appendix 2]). Note that the t_j act with degree -2.

We define the complex $\mathrm{Sh}(\mathbf{G}, \sigma)$ to be the graded free R-module

$$S\{y_1, \dots, y_c\} \otimes \mathbf{G} \otimes R,$$

with differential

$$\delta := \sum t^{\mathbf{a}} \otimes \sigma_{\mathbf{a}} \otimes R\,.$$

In this book, we often consider the case when we take only one element $f \in S$, and then we denote the divided power algebra by $S\{y\}$, where the $y^{(i)}$ form the dual basis to the basis $\{t^i\}$ of the polynomial ring $S[t]$.

Proposition 4.1.4 ([25, 55]) *Let $f_1, \dots, f_c$ be a regular sequence in a ring S, and let N be a finitely generated module over $R := S/(f_1, \dots, f_c)$. If $\mathbf{G}$ is an S-free resolution of N and σ is a system of higher homotopies for $f_1, \dots, f_c$ on $\mathbf{G}$, then* $\mathrm{Sh}(\mathbf{G}, \sigma)$ *is an R-free resolution of N.*

Remark 4.1.5 For example, the minimal free resolution of the residue field of a complete intersection in Tate's original paper [57] is obtained by applying the Shamash construction to the Koszul complex $\mathbf{G}$ with a set of higher homotopies for which $\sigma_a = 0$ for $|a| > 1$. It is called *Tate's resolution*.

4.2 The Action of the CI Operators on Ext

Corollary 4.2.1 ([25, Proposition 1.5]) *The CI operators commute up to homotopy.*

Proof Apply Proposition 4.1.2 part (4) to the map of complexes t_i. □

Construction 4.2.2 Let $f_1, \dots, f_c \in S$ be a regular sequence, set $R = S/(f_1, \dots, f_c)$, and let N be a finitely generated R-module. By part (1) of Proposition 4.1.2 the CI operators corresponding to $f_1, \dots, f_c$ on any R-free resolution of N induce maps

$$\mathrm{Ext}_R^{*+2}(N, k) \longrightarrow \mathrm{Ext}_R^{*}(N, k)\,.$$

Parts (2)–(4) of Proposition 4.1.2 show that this action is well-defined and functorial, and Corollary 4.2.1 shows that it makes $\mathrm{Ext}_R(N, k)$ into a graded module over the polynomial ring $\mathcal{R} := k[\chi_1, \cdots, \chi_c]$, where the variable χ_j acts by the dual of t_j, and thus has degree 2.

Because the χ_j have degree 2, we may split any graded $\mathcal{R}$-module into even degree and odd degree parts; in particular, we write

$$\mathrm{Ext}_R(N, k) = \mathrm{Ext}_R^{even}(N, k) \oplus \mathrm{Ext}_R^{odd}(N, k)$$

as $\mathcal{R}$-modules.

The following result shows that the action of the CI operators is highly nontrivial.

Theorem 4.2.3 ([8, 25, 31]) *Let $f_1, \ldots, f_c$ be a regular sequence in a local ring S with residue field k, and set $R = S/(f_1, \ldots, f_c)$. If N is a finitely generated R-module with finite projective dimension over S, then the action of the CI operators makes $\mathrm{Ext}_R(N, k)$ into a finitely generated $k[\chi_1, \ldots, \chi_c]$-module.*

Remark 4.2.4 Construction 4.2.2, Theorem 4.2.3, and our proof below hold verbatim for $\mathrm{Ext}_R(N, U)$ for any finitely generated R-module U.

A version of Theorem 4.2.3 was proved in [31] by Gulliksen, using a different construction of operators on Ext. Other constructions of operators were introduced and used by Avramov [4], Avramov-Sun [8], Eisenbud [25], and Mehta [42]. The relations between these constructions are explained by Avramov and Sun [8]. We will use only Construction 4.1.1. We provide a new proof of Theorem 4.2.3:

Proof Let $\mathbf{G}$ be a finite S-free resolution of N. By Proposition 3.4.2, there exists a system of higher homotopies on $\mathbf{G}$. Proposition 4.1.4 shows that $\mathrm{Sh}(\mathbf{G}, \sigma)$ is an R-free resolution of N.

By Construction 4.3.1 the CI operators can be chosen to act on $\mathrm{Sh}(\mathbf{G}, \sigma)$ as multiplication by the t_i, dual to the y_i, and thus they commute. Therefore, $\mathrm{Hom}_R(\mathrm{Sh}(\mathbf{G}, \sigma), k)$ is a finitely generated module over $\mathcal{R} := k[\chi_1, \ldots, \chi_c]$. As the CI operators commute with the differential, it follows that both the kernel and the image of the differential are submodules, so they are finitely generated as well. Thus, so is the quotient module $\mathrm{Ext}_R(N, k)$. □

Theorem 4.2.3 immediately implies numerical conditions on Betti numbers:

Theorem 4.2.5 ([31]) *Under the assumptions in Theorem 4.2.3, the Betti numbers of the R-module N are eventually given by two polynomials $P_{even}(z), P_{odd}(z) \in \mathbf{Q}[z]$ of degrees $\leq c-1$, so for $i \gg 0$:*

$$\beta^R_{2i}(N) = P_{even}(i)$$
$$\beta^R_{2i+1}(N) = P_{odd}(i)\,.$$

Its Poincaré series has the form

$$\mathcal{P}^R_N(x) = \frac{v(x)}{(1-x^2)^c}$$

for some polynomial $v(x) \in \mathbf{Z}[x]$.

Proof Since $\mathrm{Ext}^{even}_R(N, k)$ is a finitely generated graded module over the polynomial ring $k[\chi_1, \cdots, \chi_c]$, its Hilbert function is eventually given by a polynomial of degree $\leq c-1$. Similarly, the Hilbert function of $\mathrm{Ext}^{odd}_R(N, k)$ is eventually given by a polynomial of degree $\leq c-1$.

The form of the Poincaré series follows from the fact that the generating series of the polynomial ring $\mathcal{R} = k[\chi_1, \cdots, \chi_c]$ is

$$\frac{1}{(1-x^2)^c}$$

because each variable χ_j has degree 2. □

We will obtain further properties of the Betti numbers. Theorem 4.2.7 (stated somewhat differently), and the ideas of its proof are from [7, Theorem 7.3]. The proof of Lemma 4.2.6 actually does not require any machinery beyond matrix factorizations and the definition of the CI operators.

Lemma 4.2.6 *Let $f \in R'$ be a non-zerodivisor in a local ring R', and let $(\mathbf{F}, \delta)$ be a minimal free resolution over $R := R'/(f)$ of a module N. If the CI operator $t : F_2 \longrightarrow F_0$ corresponding to f is surjective, then $\beta_0^R(N) \leq \beta_1^R(N)$. Moreover, if equality holds, then $\mathbf{F}$ is periodic of period 2 and $\beta_j^R(N) = \beta_0^R(N)$ for all j.*

We will use this result again in Corollary 6.5.1 to obtain numerical information about pre-stable matrix factorizations.

Proof We lift the first two steps of $\mathbf{F}$ to R' as

$$\widetilde{F}_2 \xrightarrow{\tilde{\delta}_2} \widetilde{F}_1 \xrightarrow{\tilde{\delta}_1} \widetilde{F}_0$$

so that $\tilde{\delta}_1\tilde{\delta}_2 = f\tilde{t}$. Since t is surjective and f is in the maximal ideal, $\tilde{t}$ is surjective. Thus the image of $\tilde{\delta}_1$ contains $f\widetilde{F}_0$, and it follows that rank $(\tilde{\delta}_1)$ = rank $(\widetilde{F}_0)$. In particular, rank $(F_1) \geq$ rank (F_0).

Suppose rank (F_1) = rank (F_0). Since Coker$(\tilde{\delta}_1)$ is annihilated by f, there is a map $\tilde{u}_1 : \widetilde{F}_0 \longrightarrow \widetilde{F}_1$ such that $\tilde{\delta}_1\tilde{u}_1 = f\mathrm{Id}$. It follows from Remark 2.1.3 that $(\tilde{\delta}_1, \tilde{u}_1)$ is a matrix factorization of f. It must be minimal since if the map $\tilde{u}_1 \otimes R$ is not minimal then by minimizing the R-free resolution of N from Remark 2.1.2 we get a resolution $\mathbf{F}'$ with rank $(F_1') <$ rank (F_1), contradicting to the minimality of the resolution $\mathbf{F}$. Thus the cokernel of δ_1 is minimally resolved by the periodic resolution coming from this matrix factorization by Theorem 2.1.1, and the Betti numbers are constant. □

Theorem 4.2.7 ([7, Theorem 7.3]) *Under the assumptions in Theorem 4.2.5, for $i \gg 0$ the Betti numbers $\beta_i^R(N)$ are non-decreasing as i increases.*

Proof We may extend the residue field of R if necessary, and assume it is infinite. It then follows from primary decomposition that, for any finitely generated graded $k[\chi_1, \dots, \chi_c]$-module E, there is a linear form in $\mathcal{R} = k[\chi_1, \dots, \chi_c]$ that is a non-zerodivisor on all high truncations of E. We apply this to the module $E = \mathrm{Ext}_R(N, k)$. Any linear form in $\mathcal{R}$ corresponds to the CI operator defined by some linear combination f of the f_i. Choosing $g_1, \dots, g_{c-1}$ such that $(f_1, \dots, f_c) = (g_1, \dots, g_{c-1}, f)$ and setting $R' = S/(g_1, \dots g_{c-1})$, we may write $R = R'/(f)$ and apply Lemma 4.2.6 to all sufficiently high syzygies. □

As an immediate corollary of Theorem 4.2.7 we get:

Corollary 4.2.8 ([5, Theorem 4.1]) *The polynomials P_{even} and P_{odd} have the same degree and leading coefficient.*

Theorems 4.2.5, 4.2.7, and Corollary 4.2.8 contain nearly all that is known in general about the Betti numbers of modules over complete intersections.

Example 4.2.9 ([25]) In the case when the complete intersection R has codimension $c = 1$, Theorem 4.2.7 simply says that the Betti numbers are eventually constant. But from the theory of matrix factorizations (see Sect. 2) it follows that any maximal Cohen-Macaulay R-module without free summands has a periodic resolution, and for any R-module N the $\dim(S)$-th syzygy module of N is a maximal Cohen-Macaulay R-module without free summands. Thus the "polynomial behavior" of the Betti numbers starts after at most $\dim(R)$ steps in the resolution.

In codimension ≥ 2 however, it can take arbitrarily long for the polynomial behavior to begin. For example, suppose that R is a 0-dimensional local complete intersection R of codimension $c \geq 2$, and

$$\mathbf{F} : \cdots \longrightarrow F_1 \longrightarrow F_0 \longrightarrow k \longrightarrow 0$$

is a minimal free resolution of the residue field k of R. Since R is self-injective, the dual of an exact sequence is again exact, and $\mathrm{Hom}(k, R) = k$, so we have an exact sequence

$$\mathrm{Hom}(\mathbf{F}, R) : 0 \longrightarrow k \longrightarrow F_0^* \longrightarrow F_1^* \longrightarrow \cdots .$$

We may glue these two sequences, via the identity map $k \longrightarrow k$, to form a doubly infinite exact free complex, called the Tate resolution of k:

$$\mathbf{T} : \cdots \longrightarrow F_1 \longrightarrow F_0 \longrightarrow F_0^* \longrightarrow F_1^* \longrightarrow \cdots .$$

Truncating this we get, for every q, a free complex

$$\mathbf{T}_{\geq -q} : \cdots \longrightarrow F_1 \longrightarrow F_0 \longrightarrow F_0^* \longrightarrow F_1^* \longrightarrow \cdots \longrightarrow F_q^*,$$

and thus $\mathbf{T}_{\geq -q}[-q]$ is a minimal free resolution. Since R is 0-dimensional, it is an immediate consequence of Tate's construction of the minimal free resolution of k, described in Remark 4.1.5, that the Poincaré series of k is:

$$\mathcal{P}_k^R(x) = \frac{1}{(1-x)^c} .$$

Hence, the Betti numbers of k are given by binomial coefficients:

$$\beta_i^R(k) = \binom{i+c-1}{i}.$$

For $i \geq 0$ this agrees with the polynomial

$$P(i) := \frac{(i+c-1)\cdots(i+1)}{(c-1)!}.$$

Since $c \geq 2$ we have $P(-1) = 0$, whereas rank $(F_0^*) = 1$. This shows that the Betti numbers of $\mathbf{T}_{\geq -q}[-q]$ start to agree with a polynomial *only* after q steps, that is, starting at F_0.

4.3 Resolutions with a Surjective CI Operator

In this book we will use higher homotopies and the Shamash construction for one element $f \in S$. We focus on that case in this section.

Construction 4.3.1 Suppose that $f \in S$, and that $(\widetilde{\mathbf{G}}, \tilde{\partial})$ is a free complex over S with a system σ of higher homotopies for f, and let $\mathrm{Sh}(\widetilde{\mathbf{G}}, \sigma)$ be the corresponding Shamash complex over $R = S/(f)$. We use the notation in Construction 4.1.3.

The *standard lifting* $\widetilde{\mathrm{Sh}}(\widetilde{\mathbf{G}}, \sigma)$ of $\mathrm{Sh}(\widetilde{\mathbf{G}}, \sigma)$ to S is $S\{y\} \otimes \widetilde{\mathbf{G}}$ with the maps $\tilde{\delta} = \sum t^j \otimes \sigma_j$. In particular, $\tilde{\delta}\big|_{\widetilde{\mathbf{G}}} = \tilde{\partial}$, so of course $\tilde{\delta}^2\big|_{\widetilde{\mathbf{G}}} = \tilde{\partial}^2 = 0$. Moreover, the equations of Definition 3.4.1 say that $\tilde{\delta}^2$ acts on the complementary summand $\mathbf{G}' = \oplus_{i>0} y^{(i)}\widetilde{\mathbf{G}}$ by ft; that is, $\tilde{\delta}^2$ sends each $y^{(i)}\mathbf{G}$ isomorphically to $fy^{(i-1)}\mathbf{G}$. Thus:

$$\tilde{\delta}^2 = ft \otimes 1\,.$$

The *standard CI operator* for f on $\mathrm{Sh}(\widetilde{\mathbf{G}}, \sigma)$ is $t \otimes 1$. Note that

$$t: \ \mathrm{Sh}(\widetilde{\mathbf{G}}, \sigma) \longrightarrow \mathrm{Sh}(\widetilde{\mathbf{G}}, \sigma)[2]$$

is surjective, and is split by the map sending $y^{(i)}u \in S\{y\} \otimes \widetilde{\mathbf{G}} \otimes S/(f)$ to $y^{(i+1)}u$. Also, the *standard lifted CI operator*

$$\tilde{t} := t \otimes 1: \ \widetilde{\mathrm{Sh}}(\widetilde{\mathbf{G}}, \sigma) \longrightarrow \widetilde{\mathrm{Sh}}(\widetilde{\mathbf{G}}, \sigma)$$

commutes with the lifting $\tilde{\delta} = \sum t^j \otimes \sigma_j$ of the differential δ.

We will use the following modified version of Proposition 4.1.4:

Proposition 4.3.2 *Let $\widetilde{\mathbf{G}}$ be a complex of S-free modules with a system of higher homotopies σ for a non-zerodivisor f in a ring S. If $\mathbf{F} = \mathrm{Sh}(\widetilde{\mathbf{G}}, \sigma)$, then $H_0(\mathbf{F}) = H_0(\widetilde{\mathbf{G}})$. Moreover, for any $i > 0$ we have $H_j(\mathbf{F}) = 0$ for all $1 \leq j \leq i$ if and only if $H_j(\widetilde{\mathbf{G}}) = 0$ for all $1 \leq j \leq i$. In particular, $\mathrm{Sh}(\widetilde{\mathbf{G}}, \sigma)$ is an $S/(f)$-free resolution of a module N if and only if $\widetilde{\mathbf{G}}$ is an S-free resolution of N.*

Proof Set $R = S/(f)$. To see that $H_0(\widetilde{\mathbf{G}}) = H_0(\mathbf{F})$, note that $R \otimes \widetilde{G}_i = F_i$ for $i \leq 1$ and $H_0(\widetilde{\mathbf{G}})$ is annihilated by f.

Set $\overline{\mathbf{G}} = R \otimes \widetilde{\mathbf{G}}$. For proving exactness we use the short exact sequences of complexes

$$0 \longrightarrow \overline{\mathbf{G}} \longrightarrow \mathbf{F} \xrightarrow{t} \mathbf{F}[2] \longrightarrow 0$$

$$0 \longrightarrow \widetilde{\mathbf{G}} \xrightarrow{f} \widetilde{\mathbf{G}} \longrightarrow \overline{\mathbf{G}} \longrightarrow 0,$$

which yield long exact sequences

$$\cdots \xrightarrow{t} H_{j-1}(\mathbf{F}) \longrightarrow H_j(\overline{\mathbf{G}}) \longrightarrow H_j(\mathbf{F}) \xrightarrow{t} H_{j-2}(\mathbf{F}) \longrightarrow H_{j-1}(\overline{\mathbf{G}}) \longrightarrow \cdots \tag{4.1}$$

$$\cdots \longrightarrow H_{j+1}(\overline{\mathbf{G}}) \longrightarrow H_j(\widetilde{\mathbf{G}}) \xrightarrow{f} H_j(\widetilde{\mathbf{G}}) \longrightarrow H_j(\overline{\mathbf{G}}) \longrightarrow H_{j-1}(\widetilde{\mathbf{G}}) \longrightarrow \cdots \tag{4.2}$$

respectively. Since σ_1 is a homotopy for f on $\widetilde{\mathbf{G}}$, the latter sequence breaks up into short exact sequences

$$0 \longrightarrow H_j(\widetilde{\mathbf{G}}) \longrightarrow H_j(\overline{\mathbf{G}}) \longrightarrow H_{j-1}(\widetilde{\mathbf{G}}) \longrightarrow 0. \tag{4.3}$$

First, assume that $H_j(\mathbf{F}) = 0$ for $1 \leq j \leq i$. From the long exact sequence (4.1) we conclude that $H_j(\overline{\mathbf{G}}) = 0$ for $2 \leq j \leq i$, and then (4.3) implies that $H_j(\widetilde{\mathbf{G}}) = 0$ for $1 \leq j \leq i$.

Conversely, suppose that $H_j(\widetilde{\mathbf{G}}) = 0$ for $1 \leq j \leq i$. It is well known that if we apply the Shamash construction to a resolution then we get a resolution, but since the bound i is not usually present we give an argument:

Assume that $H_j(\widetilde{\mathbf{G}}) = 0$ for $1 \leq j \leq i$. By (4.3) it follows that $H_j(\overline{\mathbf{G}}) = 0$ for $2 \leq j \leq i$. Applying (4.1), we conclude that $H_j(\mathbf{F}) \cong H_{j-2}(\mathbf{F})$ for $3 \leq j \leq s$. Hence, it suffices to prove that $H_1(\mathbf{F}) = H_2(\mathbf{F}) = 0$.

We next prove that $H_1(\mathbf{F}) = 0$. Let $\overline{g}_1$ be a cycle in $F_1 = \overline{G}_1$, and let $g_1 \in \widetilde{G}_1$ be an element that reduces modulo f to $\overline{g}_1$. We have

$$\tilde{\partial}(g_1) = fg_0 = \tilde{\partial}\sigma_1(g_0)$$

for some $g_0 \in G_0$. Thus $g_1 - \sigma_1(g_0) \in \mathrm{Ker}(\tilde{\partial})$ is a cycle in $\widetilde{\mathbf{G}}$. Since $H_1(\widetilde{\mathbf{G}}) = 0$, we must have $g_1 - \sigma_1(g_0) = \tilde{\partial}(g_2)$ for some $g_2 \in \widetilde{G}_2$. Using the isomorphism

$\widetilde{F_2} = \widetilde{G_2} \oplus \widetilde{G_0}$ we see that

$$g_1 = \sigma_1(g_0) + \tilde{\partial}(g_2) = \tilde{\delta}(g_0 + g_2).$$

It follows that $\overline{g}_1 = \delta(\overline{g}_0 + \overline{g}_2)$ is a boundary in $\mathbf{F}$, as required.

Finally, we show that $H_2(\mathbf{F}) = 0$. Part of (4.1) is the exact sequence

$$H_2(\overline{\mathbf{G}}) \longrightarrow H_2(\mathbf{F}) \xrightarrow{t} H_0(\mathbf{F}) \xrightarrow{\beta} H_1(\overline{\mathbf{G}}) \longrightarrow H_1(\mathbf{F}) = 0.$$

Since $H_2(\overline{\mathbf{G}}) = 0$, it suffices to show that the map β (marked in the sequence above) is a monomorphism. But we already showed that $H_1(\mathbf{F}) = 0$, so β is an epimorphism. Since $H_1(\widetilde{\mathbf{G}}) = 0$, the short exact sequence (4.3) implies $H_1(\overline{G}) \cong H_0(\widetilde{\mathbf{G}}) \cong H_0(\mathbf{F})$. Since the source and target of the epimorphism β are isomorphic finitely generated modules over the ring S, we conclude that β is an isomorphism, whence $H_2(\mathbf{F}) = 0$. □

It follows from Theorem 4.2.3 that CI operators on the resolutions of high syzygies over complete intersections are often surjective, in a sense we will make precise. To prepare for the study of this situation, we consider what can be said when a CI operator is surjective.

Proposition 4.3.3 *Let $f \in S$ be a non-zerodivisor in a ring S, and let*

$$(\mathbf{F}, \delta): \quad \cdots \longrightarrow F_i \xrightarrow{\delta_i} F_{i-1} \longrightarrow \ \ldots \ \longrightarrow F_1 \xrightarrow{\delta_1} F_0$$

be a complex of free $R := S/(f)$-modules. Let $(\widetilde{\mathbf{F}}, \tilde{\delta})$ be a lifting of $(\mathbf{F}, \delta)$ to S. Set

$$\tilde{t} := (1/f)\tilde{\delta}^2 : \widetilde{\mathbf{F}} \longrightarrow \widetilde{\mathbf{F}}[2],$$
$$\widetilde{\mathbf{G}} = \mathrm{Ker}(\tilde{t}).$$

Suppose that $\tilde{t}$ is surjective. Then:

(1) [25, Theorem 8.1] *The maps $\tilde{\delta} : \widetilde{F_i} \longrightarrow \widetilde{F_{i-1}}$ induce maps*

$$\tilde{\partial} : \widetilde{G_i} \longrightarrow \widetilde{G_{i-1}}$$

and

$$\widetilde{\mathbf{G}}: \quad \cdots \longrightarrow \widetilde{G_{i+1}} \xrightarrow{\tilde{\partial}_{i+1}} \widetilde{G_i} \longrightarrow \cdots \longrightarrow \widetilde{G_1} \xrightarrow{\tilde{\partial}_1} \widetilde{G_0}$$

is an S-free complex. If S is local and $\mathbf{F}$ is minimal, then so is $\widetilde{\mathbf{G}}$.

(2) *We may write* $\widetilde{F}_i = \oplus_{j\geq 0}\widetilde{G}_{i-2j}$ *in such a way that the lifted CI operator* $\tilde{t}$ *consists of the projections*

$$\widetilde{F}_i = \bigoplus_{0\leq j\leq i/2} \widetilde{G}_{i-2j} \xrightarrow{\tilde{t}} \bigoplus_{0\leq j\leq (i-2)/2} \widetilde{G}_{i-2-2j} = \widetilde{F}_{i-2}\,.$$

If $\sigma_j : \widetilde{G}_{i-2j} \longrightarrow \widetilde{G}_{i-1}$ *denotes the appropriate component of the map* $\tilde{\delta} : \widetilde{F}_i \longrightarrow \widetilde{F}_{i-1}$, *then* $\sigma = \{\sigma_j\}$ *is a system of higher homotopies on* $\widetilde{\mathbf{G}}$, *and* $\mathbf{F} \cong \mathrm{Sh}(\widetilde{\mathbf{G}}, \sigma)$.

Proof (2): Since the maps $\tilde{t}$ are surjective, it follows inductively that we may write $\widetilde{F}_i$ and $\tilde{t}$ in the given form. The component corresponding to $\widetilde{G}_{i-2j} \longrightarrow \widetilde{G}_{i-1}$ in $\tilde{\delta} : \widetilde{F}_m \longrightarrow \widetilde{F}_{m-1}$ is the same for any m with $m \geq i-2j$ and $m \equiv i \bmod(2)$ because $\tilde{\delta}$ commutes with $\tilde{t}$. The condition that σ is a sequence of higher homotopies is equivalent to the condition that $\tilde{\delta}^2 = f\tilde{t}$, as one sees by direct computation. It is now immediate that $\mathbf{F} \cong \mathrm{Sh}(\widetilde{\mathbf{G}}, \sigma)$. □

As an immediate consequence of Propositions 4.3.2 and 4.3.3 we get:

Corollary 4.3.4 *Let* $f \in S$ *be a non-zerodivisor in a local ring* S, *and set* $R = S/(f)$. *Let* N *be an* R*-module with a minimal* R*-free resolution* $(\mathbf{F}, \delta)$ *with a surjective CI operator. Let* $(\widetilde{\mathbf{F}}, \tilde{\delta})$ *be a lifting of* $(\mathbf{F}, \delta)$ *to* S, *and set*

$$\tilde{t} := (1/f)\tilde{\delta}^2 : \widetilde{\mathbf{F}} \longrightarrow \widetilde{\mathbf{F}}[2]\,.$$

The minimal S*-free resolution of* N *is*

$$(\mathbf{G}, \partial) = \mathrm{Ker}(\tilde{t})\,.$$

If we split the epimorphisms $t : F_i \longrightarrow F_{i-2}$ *and correspondingly write* $F_i = \overline{G}_i \oplus F_{i-2}$ *(with* $\overline{G} = R \otimes \widetilde{G}$ *and* $\bar{\partial} = R \otimes \tilde{\partial}$*), then the differential* $\delta : F_i \longrightarrow F_{i-1}$ *has the form:*

$$\delta_i = \begin{array}{c} \\ \overline{G}_{i-1} \\ F_{i-3} \end{array} \begin{array}{c} \quad\overline{G}_i \quad\; F_{i-2} \\ \begin{pmatrix} \bar{\partial}_i & \varphi_i \\ O & \delta_i \end{pmatrix} \end{array}.$$

□

Also as an immediate consequence of Propositions 4.3.2 and 4.3.3 we obtain a result of Avramov-Gasharov-Peeva; their proof relies on the spectral sequence proof of [7, Theorem 4.3].

Corollary 4.3.5 ([7, Proposition 6.2]) *Let* $f \in S$ *be a non-zerodivisor in a local ring, and set* $R = S/(f)$. *Let* N *be an* R*-module with a minimal* R*-free resolution* $\mathbf{F}$. *The CI operator* χ *corresponding to* f *is a non-zerodivisor on* $\mathrm{Ext}_R(N, k)$ *if and only* *if the CI operator* $t : \mathbf{F}[-2] \longrightarrow \mathbf{F}$ *is surjective, if and only if the minimal R-free resolution of N is obtained by the Shamash construction applied to the minimal free resolution of N over S.*

Proof Nakayama's Lemma shows that the CI operator $t : \mathbf{F}[-2] \longrightarrow \mathbf{F}$ is surjective if and only if the operator $\chi : \mathrm{Ext}_R(N, k) \longrightarrow \mathrm{Ext}_R(N, k)$ is injective. □

Chapter 5
Infinite Resolutions of HMF Modules

Abstract In this chapter we construct the infinite minimal free resolution of a higher matrix factorization module.

5.1 The Minimal R-Free Resolution of a Higher Matrix Factorization Module

Let (d, h) be a higher matrix factorization with respect to a regular sequence $f_1, \ldots, f_c$ in a ring S, and $R = S/(f_1, \ldots, f_c)$. We will describe an R-free resolution of the HMF module M that is minimal when S is local and (d, h) is minimal.

Construction 5.1.1 Let (d, h) be a higher matrix factorization with respect to a regular sequence $f_1, \ldots, f_c$ in a ring S. We will adopt the notation of 1.4.1. In addition we choose splittings

$$A_s(p) = A_s(p-1) \oplus B_s(p)$$

for $s = 0, 1$, so

$$A_s(p) = \oplus_{1 \le q \le p} B_s(q) .$$

We write

$$\psi_p : B_1(p) \longrightarrow A_0(p-1)$$
$$b_p : B_1(p) \longrightarrow B_0(p)$$

for the maps induced by d. We will frequently use the two-term complexes

$$\mathbf{A}(p) : A_1(p) \longrightarrow A_0(p)$$
$$\mathbf{B}(p) : B_1(p) \xrightarrow{b_p} B_0(p) .$$

The construction is by induction on p:

D. Eisenbud, I. Peeva, *Minimal Free Resolutions over Complete Intersections*, Lecture Notes in Mathematics 2152, DOI 10.1007/978-3-319-26437-0_5

- Let $p = 1$. Set $\mathbf{U}(1) = \mathbf{B}(1)$, and note that h_1 is a homotopy for f_1. Set

$$\mathbf{T}(1) := \mathrm{Sh}(\mathbf{U}(1), h_1)$$

Its first two terms are the complex $R(1) \otimes \mathbf{B}(1) = R(1) \otimes \mathbf{A}(1)$.
- For $p \geq 2$, given an $R(p-1)$-free resolution $\mathbf{T}(p-1)$ of $M(p-1)$ with beginning $R(p-1) \otimes \mathbf{A}(p-1)$, let

$$\Psi_p : R(p-1) \otimes \mathbf{B}(p)[-1] \longrightarrow \mathbf{T}(p-1)$$

be the map of complexes induced by $\psi_p : B_1(p) \longrightarrow A_0(p-1)$. Set

$$\mathbf{U}(p) := \mathbf{Cone}\left(\Psi_p\right) ,$$

as shown in the diagram:

$$\begin{array}{ccccccc} \mathbf{T}(p-1): \cdots \to T_3(p-1) \longrightarrow T_2(p-1) \to & A_1'(p-1) & \xrightarrow{d_{p-1}'} & A_0'(p-1) \\ & \oplus & \overset{\psi_p'}{\nearrow} & \oplus \\ & B_1'(p) & \xrightarrow[b_p']{} & B_0'(p), \end{array}$$

where $-'$ denotes $R(p-1) \otimes -$.

We will show that $\mathbf{U}(p)$ is an $R(p-1)$-free resolution of $M(p)$. Thus we can choose a system of higher homotopies $\sigma(p)$ for f_p on $\mathbf{U}(p)$ beginning with

$$\sigma(p)_1 := R(p-1) \otimes h_p : R(p-1) \otimes A_0(p) \longrightarrow R(p-1) \otimes A_1(p).$$

Note that $\sigma(p)_0$ is the differential d_p. Set

$$\mathbf{T}(p) := \mathrm{Sh}(\mathbf{U}(p), \sigma(p)).$$

The underlying graded module of $\mathbf{T}(p)$ is $\mathbf{U}(p) = \mathbf{Cone}(\Psi_p)$ tensored with a divided power algebra on a variable y_p of degree 2. Its first differential is

$$R(p) \otimes \mathbf{A}(p) : R(p) \otimes A_1(p) \xrightarrow{R(p) \otimes d_p} R(p) \otimes A_0(p),$$

which is the presentation of $M(p)$. We see by induction on p that the term $T_j(p)$ of homological degree j in $\mathbf{T}(p)$ is a direct sum of the form

$$T_j(p) = \bigoplus y_{q_1}^{(a_1)} \cdots y_{q_i}^{(a_i)} B_s(q) \otimes R(p) , \tag{5.1}$$

where the sum is over all terms with

$$\begin{aligned}0 &\leq s \leq 1,\\ p &\geq q_1 > q_2 > \cdots > q_i \geq q \geq 1,\\ a_m &> 0 \text{ for } 1 \leq m \leq i,\\ j &= s + \sum_{1\leq m\leq i} 2a_m.\end{aligned}$$

We say that such an element $y_{q_1}^{(a_1)} \cdots y_{q_i}^{(a_i)} v$ with $0 \neq v \in B_s(q)$ is *admissible* and has *weight* q_1 if $a_1 > 0$, and we make the convention that the admissible elements in $B_s(q)$ have weight 0.

The complex $\mathbf{T}(c)$ is thus filtered by:

$$\mathbf{T}(0) := 0 \subseteq R \otimes \mathbf{T}(1) \subseteq \cdots \subseteq R \otimes \mathbf{T}(p-1) \subseteq \mathbf{T}(c),$$

where $R \otimes \mathbf{T}(p)$ is the subcomplex spanned by elements of weight $\leq p$ with $v \in B_s(q)$ for $q \leq p$.

Theorem 5.1.2 *With notation and hypotheses as in Construction* 5.1.1*:*

(1) *The complex* $\mathbf{T}(p)$ *is an* $R(p)$*-free resolution of* $M(p)$ *whose first differential is* $R(p) \otimes d_p$ *and whose second differential is*

$$R(p) \otimes \Big(\Big(\oplus_{q\leq p} A_0(q) \Big) \xrightarrow{h} A_1(p) \Big),$$

where the q*-th component of* h *is*

$$h_q : A_0(q) \longrightarrow A_1(q) \hookrightarrow A_1(p).$$

(2) *If* S *is local then* $\mathbf{T}(p)$ *is the minimal free resolution of* $M(p)$ *if and only if the higher matrix factorization* $\big(d_p, h(p) = (h_1|\cdots|h_p)\big)$ *(see* 1.4.1 *for notation) is minimal.*

Proof of Theorem 5.1.2(1) We do induction on p. To start the induction, note that $\mathbf{U}(1)$ is the two-term complex $\mathbf{A}(1) = \mathbf{B}(1)$. By hypothesis, its differential d_1 and homotopy h_1 form a hypersurface matrix factorization for f_1, and $\mathbf{T}(1)$ has the form:

$$\mathbf{T}(1) : R(1) \otimes \Big(\cdots \xrightarrow{h_1} A_1(1) \xrightarrow{d_1} A_0(1) \xrightarrow{h_1} A_1(1) \xrightarrow{d_1} A_0(1) \Big).$$

Inductively, suppose that $p \geq 2$, and that

$$\mathbf{T}(p-1) : \cdots \longrightarrow T_2 \longrightarrow T_1 \longrightarrow T_0$$

is an $R(p-1)$-free resolution of $M(p-1)$ whose first two maps are as claimed. We write $\overline{}$ for $R(p-1) \otimes -$. It follows that the first map of $\mathbf{U}(p)$ is

$$\overline{d}(p) : \overline{A}_1(p) = T_1 \oplus \overline{B}_1(p) \longrightarrow \overline{A}_0(p) = T_0 \oplus \overline{B}_0(p).$$

Since $R(p-1) \otimes \big(d_p h_p\big) = f_p \operatorname{Id}_{A_0(p)}$ we may take $R(p-1) \otimes h_p$ to be the start of a system of higher homotopies $\sigma(p)$ for f_p on $R(p-1) \otimes \mathbf{U}(p)$. It follows from the definition that the first two maps in $\mathbf{T}(p) = \operatorname{Sh}\big(\mathbf{U}(p), \sigma(p)\big)$ are as asserted.

By Proposition 4.3.2, the Shamash construction takes an $R(p-1)$-free resolution to an $R(p)$-free resolution of the same module. Thus for the induction it suffices to show that $\mathbf{U}(p)$ is an $R(p-1)$-free resolution of $M(p)$. Since the first map of $\mathbf{U}(p)$ is $\overline{d}(p)$, and since $\overline{h}(p)$ is a homotopy for f_p, we see at once that

$$H_0(\mathbf{U}(p)) = \operatorname{Coker}(\overline{d}(p)) = \operatorname{Coker}(R(p) \otimes d_p) = M(p).$$

To prove that $\mathbf{U}(p)$ is a resolution, note first that

$$\mathbf{U}(p)_{\geq 2} = \mathbf{T}(p-1)_{\geq 2},$$

and the image of $U(p)_2 = T(p-1)_2$ is contained in the summand $T(p-1)_1 \subseteq U(p)_1$, so

$$H_i(\mathbf{U}(p)) = H_i(\mathbf{T}(p-1)) = 0$$

for $i \geq 2$. Thus it suffices to prove that $H_1(\mathbf{U}(p)) = 0$.

Let

$$(y, v) \in U(p)_1 = T(p-1)_1 \oplus \overline{B}(p)_1$$

be a cycle in $\mathbf{U}(p)$. Thus, $\overline{b}_p(v) = 0$ and $\overline{\psi}_p(v) = -\overline{d}_{p-1}(y)$. By Lemma 3.1.7, we conclude that $v = 0$. □

For the proof of part (2) of Theorem 5.1.2 we will use the form of the resolutions $\mathbf{T}(p)$ to make a special lifting of the differentials to S, and thus to produce especially "nice" CI operators. We pause in the proof of Theorem 5.1.2 to describe this construction and deduce some consequences.

Proposition 5.1.3 *With notation and hypotheses as in Construction* 5.1.1, *there exists a lifting of the filtration*

$$\mathbf{T}(1) \subseteq \cdots \subseteq \mathbf{T}(c)$$

to a filtration

$$\widetilde{\mathbf{T}}(1) \subseteq \cdots \subseteq \widetilde{\mathbf{T}}(c)$$

over S, and a lifting $\tilde{\delta}$ of the differential δ in $\mathbf{T}(c)$ to S with lifted CI operators $\tilde{t}_1, \dots, \tilde{t}_c$ on $\widetilde{\mathbf{T}}(c)$ such that for every $1 \le p \le c$:

(1) *Both $\tilde{\delta}$ and $\tilde{t}_p$ preserve $\widetilde{\mathbf{T}}(p)$, and $\tilde{t}_p\big|_{\widetilde{\mathbf{T}}(p)}$ commutes with $\tilde{\delta}\big|_{\widetilde{\mathbf{T}}(p)}$ on $\widetilde{\mathbf{T}}(p)$.*

(2) *The CI operator t_p vanishes on the subcomplex $R \otimes \mathbf{U}(p)$ and induces an isomorphism from $R \otimes T(p)_j/U(p)_j$ to $R \otimes T(p)_{j-2}$ that sends an admissible element $y_{q_1}^{(a_1)} \cdots y_{q_i}^{(a_i)} v$ with $q_1 = p$ to $y_{q_1}^{(a_1-1)} \cdots y_{q_i}^{(a_i)} v$.*

Proof The proof is by induction on p. If $p = 1$ the result is obvious. Thus we may assume by induction that liftings

$$0 \subset \widetilde{\mathbf{T}}(1) \subseteq \cdots \subseteq \widetilde{\mathbf{T}}(p-1),$$

$\tilde{\delta}(p-1)$ and $\tilde{t}_1, \dots, \tilde{t}_{p-1}$ on $\widetilde{\mathbf{T}}(p-1)$ satisfying the Proposition have been constructed. We use the maps ψ_p and b_p from the definition of the higher matrix factorization to construct a lifting of $\mathbf{U}(p)$ from the given lifting of $\mathbf{T}(p-1)$. In addition, we choose liftings $\tilde{\sigma}$ of the maps (other than the differential) in the system of higher homotopies $\sigma(p)$ for f_p on $\mathbf{U}(p)$.

By construction, $\mathbf{T}(p) = \mathrm{Sh}(\mathbf{U}(p), \sigma(p))$, so we take the standard lifting to S from Construction 4.3.1, that is, take

$$\widetilde{\mathbf{T}}(p) = \oplus_{i \ge 0}\, y_p^{(i)} \widetilde{\mathbf{U}}(p)$$

with lifting of the differential $\tilde{\delta} = \sum t^j \otimes \tilde{\sigma}_j$, where t is the dual variable to y_p.

By Construction 4.3.1 it follows that, modulo $(f_1, \dots, f_{p-1})$, the map $\tilde{\delta}^2$ vanishes on $\widetilde{\mathbf{U}}(p)$ and induces f_p times the projection $\widetilde{T}_j(p)/\widetilde{U}_j(p) \longrightarrow \widetilde{T}_{j-2}(p)$.

We choose $\tilde{t}_p$ to be the standard lifted CI operator, which vanishes on $\widetilde{\mathbf{U}}(p)$ and is the projection $\widetilde{T}_j(p)/\widetilde{U}_j(p) \longrightarrow \widetilde{T}_{j-2}(p)$. Then $\tilde{\delta}_{i-2}\tilde{t}_p = \tilde{t}_p\tilde{\delta}_i$ by construction; see Construction 4.3.1.

Recall that $\tilde{\delta}\big|_{\widetilde{\mathbf{T}}(p-1)}$ is the lifting $\tilde{\delta}(p-1)$ given by induction. Therefore, from $\tilde{\delta}$ we can choose maps $\tilde{t}_1, \dots, \tilde{t}_{p-1}$ on $\widetilde{\mathbf{T}}(p)$ that extend the maps $\tilde{t}_1, \dots, \tilde{t}_{p-1}$ given by induction on $\widetilde{\mathbf{T}}(p-1) \subseteq \widetilde{\mathbf{U}}(p)$. □

The CI operators on a minimal R-free resolution commute up to homotopy by Corollary 4.2.1. It was conjectured in [25] that they could be chosen to commute, and even that the resolution could be embedded in the Shamash resolution (on which the standard CI operators obviously commute). An obstruction to this embedding is given in [7, Theorem 9.1], and [7, Example 9.3] provides a counterexample. However, this leaves open the conjecture for a high truncation of the minimal free resolution. Avramov and Buchweitz [6] showed that the embedding is indeed possible for high syzygies in codimension 2. A recent counterexample in [29] shows that it is not possible to choose commuting CI operators in codimension 3, even for arbitrarily high syzygies. Thus the following result is sharp:

Theorem 5.1.4 *Suppose that S is local. With CI operators on $\mathbf{T}(p)$ chosen as in Proposition* 5.1.3, *the operator t_p commutes on $\mathbf{T}(p)$ with each t_i for $i < p$.*

Proof The theorem follows easily from Proposition 5.1.3 together with the following general criterion in Lemma 5.1.5. □

Lemma 5.1.5 *Let $f_1, \dots, f_c$ be a regular sequence in a local ring S, and let $R = S/(f_1, \dots, f_c)$. Suppose that $(\mathbf{F}, \delta)$ is a complex over R with lifting $(\widetilde{\mathbf{F}}, \tilde{\delta})$ to S, and let $\tilde{t}_1, \dots, \tilde{t}_c$ on $\widetilde{\mathbf{F}}$ define CI operators corresponding to $f_1, \dots, f_c$. If, for some j, $\widetilde{t_j}$ commutes with $\tilde{\delta}^2$, then t_j commutes with each t_i.*

Proof Since $\tilde{\delta}^2 = \sum f_i \tilde{t}_i$ by definition, we have

$$\sum f_i \tilde{t}_j \tilde{t}_i = \sum f_i \tilde{t}_i \tilde{t}_j\,,$$

or equivalently

$$\sum f_i(\tilde{t}_j \tilde{t}_i - \tilde{t}_i \tilde{t}_j) = 0\,.$$

Since $f_1, \dots, f_c$ is a regular sequence it follows that $\tilde{t}_j \tilde{t}_i - \tilde{t}_i \tilde{t}_j$ is zero modulo $(f_1, \dots, f_c)$ for each i. □

We will now complete the proof of Theorem 5.1.2:

Proof of Theorem 5.1.2(2) We suppose that S is local with maximal ideal $\mathbf{m}$. If the resolution $\mathbf{T}(p)$ is minimal then it follows at once from the description of the first two maps that (d, h) is minimal. We will prove the converse by induction on p.

If $p = 1$ then $\mathbf{T}(1)$ is the periodic resolution

$$\mathbf{T}(1): \quad \cdots \xrightarrow{h_1} A_1 \xrightarrow{d_1} A_0 \xrightarrow{h_1} A_1 \xrightarrow{d_1} A_0$$

and only involves the maps (d_1, h_1); this is obviously minimal if and only if d_1 and h_1 are minimal.

Now suppose that $p > 1$ and that $\mathbf{T}(q)$ is minimal for $q < p$. Let $\delta_i : T_i(p) \longrightarrow T_{i-1}(p)$ be the differential of $\mathbf{T}(p)$. We will prove minimality of δ_i by a second induction, on i, starting with $i = 1, 2$.

Recall that the underlying graded module of $\mathbf{T}(p) = \mathrm{Sh}(\mathbf{U}(p), \sigma)$ is the divided power algebra $S\{y_p\} = \sum_i S y_p^{(i)}$ tensored with the underlying module of $R(p) \otimes \mathbf{U}(p)$. Thus the beginning of the resolution $\mathbf{T}(p)$ has the form

$$\cdots \longrightarrow R(p) \otimes y_p A_0(p) \oplus R(p) \otimes T_2(p-1) \xrightarrow{\delta_2} R(p) \otimes A_1(p) \xrightarrow{\delta_1} R(p) \otimes A_0(p).$$

The map δ_1 is induced by d_p, which is minimal by hypothesis. Further, $\delta_2 = (h_p, \partial_2)$ where the map ∂_2 is the differential of $\mathbf{T}(p-1)$ tensored with $R(p)$. The map h_p is minimal by hypothesis, and ∂ is minimal by induction on p, so δ_2 is minimal as well.

Now suppose that $j \geq 2$ and that δ_i is minimal for $i \leq j$. We must show that δ_{j+1} is minimal, that is, $\delta_{j+1}(w) \in \mathbf{m}T_j(p)$ for any $w \in T_{j+1}(p)$. By Construction 5.1.1, $\delta_{j+1}(w)$ can be written uniquely as a sum of admissible elements of the form

$$y_{q_1}^{(a_1)} \cdots y_{q_i}^{(a_i)} v.$$

Admissibility means that $0 \neq v \in B_s(q)$ and

$$\begin{aligned} &0 \leq s \leq 1, \\ &p \geq q_1 > q_2 > \cdots > q_i \geq q \geq 1, \\ &a_m > 0 \ \text{ for } 1 \leq m \leq i, \\ &j = s + \sum_{1 \leq m \leq i} 2a_m . \end{aligned}$$

If $\delta_{j+1}(w) \notin \mathbf{m}T_j(p)$ then there exists a summand $y_{q_1}^{(a_1)} \cdots y_{q_i}^{(a_i)} v$ in this expression that is not in $\mathbf{m}T_j(p)$. Since $\delta_{j+1}(w)$ has homological degree $j \geq 2$, the weight of this summand must be > 0, that is, a factor $y_{q_1}^{(a_1)}$ must be present.

Choose such a summand with weight q_1' as large as possible. We choose $t_{q_1'}$ as in Proposition 5.1.3. The map $t_{q_1'}$ sends every admissible element of weight $< q_1'$ to zero. The admissible summands of $\delta_{j+1}(w)$ with weight $> q_1'$ can be ignored since they are in $\mathbf{m}T_{j-2}(p)$. By Proposition 5.1.3 it follows that $t_{q_1'}\delta_{j+1}(w) \notin \mathbf{m}T_{j-2}(p)$. Since

$$t_{q_1'}\delta_{j+1}(w) = \delta_{j-1}t_{q_1'}(w) ,$$

this contradicts the induction hypothesis. □

The filtration of the resolution defined by Theorem 5.1.2 gives a simple filtration of the Ext module:

Corollary 5.1.6 *Let* $k[\chi_1, \ldots, \chi_c]$ *act on* $\operatorname{Ext}_R(M, k)$ *as in Construction* 4.2.2. *There is an isomorphism*

$$\operatorname{Ext}_R(M, k) \cong \bigoplus_{p=1}^{c} k[\chi_p, \ldots, \chi_c] \otimes_k \operatorname{Hom}_S(\mathbf{B}(p), k)$$

of vector spaces such that, for $i \geq p$, χ_i *preserves the summand*

$$k[\chi_p, \ldots, \chi_c] \otimes \operatorname{Hom}_S(\mathbf{B}(p), k)$$

and acts on it via the action on the first factor.

Proof Since $\mathbf{T}(c)$ is a minimal free resolution of M, the $k[\chi_1,\ldots,\chi_c]$-module $\mathrm{Ext}_R(M,k)$ is isomorphic to $\mathrm{Hom}_R(\mathbf{T}(c),k)$. Using the decomposition in (5.1) we see that the underlying graded free module of $\mathrm{Hom}_R(\mathbf{T}(c),k)$ is

$$\bigoplus_p k[\chi_p,\ldots,\chi_c]\otimes_k \mathrm{Hom}_S(\mathbf{B}(p),k).$$

From part (2) of Proposition 5.1.3 we see that, for $i\geq p$, the action of χ_i on the summand $k[\chi_p,\ldots,\chi_c]\otimes_k\mathrm{Hom}_S(\mathbf{B}(p),k)$ is via the natural action on the first factor. □

Corollary 5.1.6 provides a standard decomposition of $\mathrm{Ext}_R(M,k)$ in the sense of Eisenbud and Peeva [27].

5.2 Betti Numbers

We can make the results on Betti numbers in Theorem 4.2.5 and Corollary 4.2.8 more explicit. Throughout this section we assume that S is local and that (d,h) is a minimal HMF, with notations and hypotheses as in Construction 5.1.1.

Corollary 5.2.1

(1) *The Poincaré series of M over R is*

$$\mathcal{P}^R_M(x)=\sum_{1\leq p\leq c}\frac{1}{(1-x^2)^{c-p+1}}\Big(x\,\mathrm{rank}\,(B_1(p))+\mathrm{rank}\,(B_0(p))\Big).$$

(2) *The Betti numbers of M over R are given by the following two polynomials in z:*

$$\beta^R_{2z}(M)=\sum_{1\leq p\leq c}\binom{c-p+z}{c-p}\mathrm{rank}\,(B_0(p))$$

$$\beta^R_{2z+1}(M)=\sum_{1\leq p\leq c}\binom{c-p+z}{c-p}\mathrm{rank}\,(B_1(p)).$$

Proof For (2), recall that the Hilbert function of $k[Z_p,\ldots,Z_c]$ is:

$$g_p(z)=\binom{c-p+z}{c-p+1}.$$

□

Recall that the *complexity* of an R-module N is defined to be

$$\mathrm{cx}_R(N) = \inf\left\{q \geq 0 \,\middle|\, \text{there exists a } w \in \mathbf{R} \text{ such that } \beta_i^R(N) \leq w i^{q-1} \text{ for } i \gg 0\right\}.$$

If the complexity of N is μ then, as noted above,

$$\dim_k \mathrm{Ext}_R^{2i}(N,k) = (\beta/(\mu-1)!)i^{\mu-1} + O(i^{\mu-2})$$

for $i \gg 0$. Following [6, 7.3] β is called the *Betti degree* of N and denoted $\mathrm{Bdeg}(N)$; this is the multiplicity of the module $\mathrm{Ext}_R^{even}(N,k)$, which is equal to the multiplicity of the module $\mathrm{Ext}_R^{odd}(N,k)$.

Corollary 5.2.2 *If (d,h) and (d',h') are two minimal higher matrix factorizations of the same R-module M (possibly with respect to different regular sequences $f_1,\dots,f_c$ and $f'_1,\dots,f'_c$), then for each $1 \leq p \leq c$:*

$$\mathrm{rank}\left(B_0(p)\right) = \mathrm{rank}\left(B'_0(p)\right)$$
$$\mathrm{rank}\left(B_1(p)\right) = \mathrm{rank}\left(B'_1(p)\right).$$

Proof First, compare the leading coefficients of the polynomials giving the Betti numbers in Corollary 5.2.1(2). Then compare the next coefficients, and so on. □

Corollary 5.2.3 *Set*

$$\gamma = \min\{\, p \mid B_1(p) \neq 0\,\}.$$

The complexity of $M = M(c)$ is

$$\mathrm{cx}_R\, M = c - \gamma + 1\,,$$

and γ is the minimal number such that $A(\gamma) \neq 0$. The Betti degree of M is

$$\mathrm{Bdeg}(M) = \mathrm{rank}\,(B_1(\gamma)) = \mathrm{rank}\,(B_0(\gamma))\,.$$

Furthermore,

$$\mathrm{rank}\,(B_1(p)) > 0 \quad \textit{for every } \gamma \leq p \leq c\,.$$

Proof This is an immediate consequence of Corollary 3.2.7. The expression of the Betti degree follows from Corollary 5.2.1(1). □

5.3 Strong Matrix Factorizations

We consider a stronger version of Definition 1.2.2 in which the map h is part of a homotopy. In Theorem 5.3.2 we show that an HMF module always has a strong matrix factorization.

Recall the Definition 1.2.3 of a strong matrix factorization. First, we remark that condition (a′) in Definition 1.2.3 is stronger than condition (a) in Definition 1.2.2. For example, in the codimension 2 case, see Example 3.1.6 and (3.1), a higher matrix factorization satisfies

$$dh_2 \equiv f_2\mathrm{Id} \mod (f_1 B_0(1) \oplus f_1 B_0(2))$$

on $B_0(1) \oplus B_0(2)$, whereas a strong matrix factorization satisfies

$$dh_2 \equiv f_2\mathrm{Id} \mod (f_1 B_0(2)) .$$

The following result gives a more conceptual description of condition (a′) and an existence statement.

Theorem 5.3.1 *Let* (d, h) *be a higher matrix factorization for a regular sequence* $f_1, \dots, f_c \in S$, *and let* $M = \mathrm{Coker}(R \otimes d)$ *be its corresponding HMF module. For* $p = 1, \dots, c$ *let* $\mathbf{L}(p)$ *be the S-free resolution of* $M(p)$ *described in Construction* 3.1.3. *With notation as in Construction* 3.1.3, *we have splittings so that*

$$A_s(p) = \bigoplus_{q=1}^{p} B_s(q)$$

for all p *and* $s = 0, 1$. *Let*

$$g : \bigoplus_{q=1}^{c} A_0(q) \longrightarrow A_1$$

be a map preserving filtrations; thus it has components

$$\bigoplus_{q=1}^{p} B_0(q) \xrightarrow{g_p} \bigoplus_{q=1}^{p} B_1(q) .$$

The pair (d, g) *is a strong matrix factorization for* $f_1, \dots, f_c \in S$ *if and only if each* g_p *can be extended to a homotopy for* f_p *on the free resolution* $\mathbf{L}(p)$.

In particular, there exists a strong matrix factorization (d, g) *of* M.

Proof We make use of the notation of Construction 3.1.3.

Condition (a′) in Definition 1.2.3 is equivalent to the property that g_p can be extended to a homotopy for f_p on $\mathbf{L}(p)_0$. Every such homotopy can be extended to a homotopy on $\mathbf{L}(p)$.

Thus it suffices to show that if each g_p can be extended to a homotopy $\sigma(p)$ for f_p on the free resolution $\mathbf{L}(p)$, then (d, g) satisfies condition (b) in Definition 1.2.2. Fix p, and denote ∂ the differential in $\mathbf{L}(p)$. Thus,

$$f_p \mathrm{Id}_{B_1(p)} = \pi_p\Big(f_p \mathrm{Id}_{A_1(p)}\Big) = \pi_p\Big(\sigma(f_p)\partial + \partial\sigma(f_p)\Big)\Big|_{A_1(p)}.$$

The second differential in $\mathbf{L}(p)$ is mapping

$$\begin{array}{l} L(p)_2 = \Big(\oplus_{i<q\leq p}\ e_i B_1(q)\Big) \oplus \Big(\oplus_{j<i<q\leq p}\ e_i e_j B_0(q)\Big) \\ \qquad\downarrow \\ L(p)_1 = \Big(\oplus_{q\leq p}\ B_1(q)\Big) \oplus \Big(\oplus_{i<q\leq p}\ e_i B_0(q)\Big). \end{array}$$

By Remark 3.1.5 the only components of the differential in $\mathbf{L}(p)$ that land in $B_1(p)$ are

$$f_i : e_i B_1(p) \longrightarrow B_1(p) \quad \text{for } i < p\,.$$

Denote $\sigma(f_p)_{e_i B_1(p)\leftarrow A_1(p)}$ the component of the homotopy $\sigma(f_p)$ with source $A_1(p)$ and target $e_i B_1(p)$. Therefore,

$$\begin{aligned} f_p \mathrm{Id}_{B_1(p)} &= \pi_p \sigma(f_p)\partial\big|_{A_1(p)} + \pi_p \partial\sigma(f_p)\big|_{A_1(p)} \\ &= \pi_p g_p d_p + \sum_{1\leq i<p} f_i \sigma(f_p)_{e_i B_1(p)\leftarrow A_1(p)} \\ &\equiv \pi_p g_p d_p \bmod (f_1, \ldots, f_{p-1}) B_1(p)\,. \end{aligned}$$

□

The next result shows that converting an HMF to a strong HMF does not destroy minimality.

Theorem 5.3.2 *Let (d, h) be a higher matrix factorization for a regular sequence $f_1, \ldots, f_c \in S$, and $M = \mathrm{Coker}(R \otimes d)$. If the ring S is local and the higher matrix factorization (d, h) is minimal, then every strong matrix factorization (d, g), having the same filtrations, is minimal as well.*

Proof We have to prove that the map g is minimal. By Theorem 5.1.2, (d, h) yields a minimal R-free resolution $\mathbf{T}$ of the module M. Again by Theorem 5.1.2, (d, g) yields an R-free resolution $\mathbf{T}'$ of M. Both resolutions have the same ranks of the corresponding free modules in them because the free modules in the filtrations of

(d, h) and (d, g) have the same ranks. Therefore, the resolution $\mathbf{T}'$ is minimal as well. The second differential in $\mathbf{T}'$ is $g \otimes R$. Hence, the map g is minimal. □

5.4 Resolutions over Intermediate Rings

Using a slight extension of the definition of a higher matrix factorization we can describe the minimal free resolutions of the modules $M(p)$ over any of the rings $R(q)$ with $q < p$, in particular the minimal free resolution of M over any $R(q)$.

Definition 5.4.1 A *generalized matrix factorization* over a ring S with respect to a regular sequence $f_1, \ldots, f_c \in S$ is a pair of maps (d, h) satisfying the definition of a higher matrix factorization *except* that we drop the assumption that $A(0) = 0$, so that we have a map of free modules $A_1(0) \xrightarrow{b_0} A_0(1)$. We do *not* require the existence of a map h_0.

Construction 5.4.2 Let (d, h) be a generalized matrix factorization with respect to a regular sequence $f_1, \ldots, f_c$ in a ring S. Using notation as in 1.4.1, we choose splittings

$$A_s(p) = A_s(p-1) \oplus B_s(p)$$

for $s = 0, 1$, and write ψ_p for the component of d_p mapping $B_1(p)$ to $A_0(p-1)$.

- Let $\mathbf{V}$ be a free resolution of the module $\mathrm{Coker}(b_0)$ over S, and set

$$\mathbf{Q}(0) := \mathbf{V}\,.$$

- Let

$$\Psi_1 : \mathbf{B}(1)[-1] \longrightarrow \mathbf{Q}(0)$$

  be the map of complexes induced by $\psi_1 : B_1(1) \longrightarrow A_0(0)$, and set

$$\mathbf{Q}(1) = \mathbf{Cone}(\Psi_1).$$

- For $p \geq 2$, suppose that an S-free resolution $\mathbf{Q}(p-1)$ of $M(p-1)$ with first term

$$Q_0(p-1) = A_0(p-1)$$

  has been constructed. Let

$$\psi'_p : \mathbf{B}(p)[-1] \longrightarrow \mathbf{L}(p-1)$$

be the map of complexes induced by $\psi_p : B_1(p) \longrightarrow A_0(p-1)$, and let

$$\Psi_p : \mathbf{K}(f_1, \dots, f_{p-1}) \otimes \mathbf{B}(p)[-1] \longrightarrow \mathbf{Q}(p-1)$$

be an $(f_1, \dots, f_{p-1})$-Koszul extension. Set $\mathbf{Q}(p) = \mathbf{Cone}(\Psi_p)$.

The proof of Theorem 3.1.4 can be applied in this situation and yields the following result.

Proposition 5.4.3 *Let (d, h) be a generalized matrix factorization over a ring S, and let $\mathbf{V}$ be a free resolution of the module* Coker(b_0) *over S. For each p, the complex $\mathbf{Q}(p)$, constructed in* 5.4.2, *is an S-free resolution of the module $M(p)$. If the ring S is local then the resulting free resolution is minimal if and only if (d, h) and $\mathbf{V}$ are minimal.* □

Theorem 5.4.4 *Let (d, h) be a higher matrix factorization. Fix a number $1 \leq j \leq c-1$. Let $\mathbf{T}(j)$ be the free resolution of $M(j)$ over the ring*

$$R(j) = S/(f_1, \dots, f_j)$$

given by Theorem 5.1.1. *Let (d', h') be the generalized matrix factorization over the ring $R(j)$ with*

$$A_s(0) = R(j) \otimes \Big(\oplus_{1 \leq q \leq j} A_s(q) \Big) \quad \text{and} \quad d'_0 = R(j) \otimes d_j,$$

$$\text{for } p > j, \;\; A_s(p)' = R(j) \otimes A_s(p+j) \quad \text{and} \quad d'_p = R(j) \otimes d_{p+j},$$

for $s = 0, 1$ and maps induces by (d, h). Then $M'(0) = M(j)$.

1. *Construction* 5.4.2, *starting from the $R(j)$-free resolution $\mathbf{Q}(0) := \mathbf{T}(j)$ of $M'(0) = M(j)$, produces a free resolution $\mathbf{Q}(c-j)$ of M over $R(j)$.*
2. *If S is local and (d, h) is minimal, then the resolution $\mathbf{Q}(c-j)$ is minimal. In that case, the Poincaré series of M over $R(j)$ is*

$$\mathcal{P}_M^{R(j)}(x) = \left(\sum_{1 \leq p \leq j} \frac{1}{(1-x^2)^{p-j-1}} \Big(x \,\text{rank}\,(B_1(p)) + \text{rank}\,(B_0(p)) \Big) \right)$$

$$\left(\sum_{j+1 \leq p \leq c} (1+x)^{p-j-1} \Big(x \,\text{rank}\,(B_1(p)) + \text{rank}\,(B_0(p)) \Big) \right).$$

Proof First, we apply Theorem 5.1.1, which gives the resolution $\mathbf{T}(j)$ of $M(j)$ over the ring $R(j)$. Then we apply Proposition 5.4.3. □

Corollary 5.4.5 *Set $\gamma = \min\{p \mid B_1(p) \neq 0\}$, so $cx_R(M) = c - \gamma + 1$ by Corollary 5.2.3. For every $j \leq \gamma - 1$, the projective dimension of M over $R(j)$ is finite and we have the equality of Poincaré series:*

$$\mathcal{P}_M^S(x) = (1+x)^j \mathcal{P}_M^{R(j)}(x) \ .$$

Chapter 6
Far-Out Syzygies

Abstract In this chapter we prove that every high syzygy over a complete intersection is a higher matrix factorization module.

6.1 Pre-stable Syzygies and Generic CI Operators

In this section we introduce the concepts of *pre-stable syzygy* and *stable syzygy* over a local complete intersection. We will see in Theorem 6.1.2 and (in a more explicit form) in Theorem 6.1.7 that any sufficiently high syzygy in a minimal free resolution over a local complete intersection ring is a stable syzygy. In Sect. 6.4 we will show that every such syzygy is an HMF module.

Definition 6.1.1 Suppose that $f_1, \dots, f_c$ is a regular sequence in a local ring S, and set $R = S/(f_1, \dots, f_c)$. We define the concepts of pre-stable and stable syzygy recursively: We say that an R-module M of finite projective dimension over S is a *pre-stable syzygy* with respect to $f_1, \dots, f_c$ if either $c = 0$ and $M = 0$, or $c \geq 1$ and M is the second syzygy over R of some R-module L of finite projective dimension over S such that:

(1) The CI operator t_c is surjective on the minimal free resolution of L;
(2) The second syzygy $\widetilde{M}$ of L as an $\widetilde{R} := S/(f_1, \dots, f_{c-1})$ module is a pre-stable syzygy with respect to $f_1, \dots, f_{c-1}$.

We say that a pre-stable syzygy is *stable* if S is a Cohen-Macaulay ring, L is a maximal Cohen-Macaulay module, and the module $\widetilde{M}$ in Condition (2) is a stable syzygy.

The goal of this section is to prove:

Theorem 6.1.2 *If $f_1, \dots, f_c \in S$ is a regular sequence in a regular local ring S, and $R = S/(f_1, \dots, f_c)$ then every sufficiently high syzygy over R is a stable syzygy with respect to any sufficiently general choice of generators for the ideal $(f_1, \dots, f_c)$.*

The definition of a stable syzygy is best understood in the context when R is a Gorenstein ring and M is a maximal Cohen-Macaulay R-module—this will always

D. Eisenbud, I. Peeva, *Minimal Free Resolutions over Complete Intersections*, Lecture Notes in Mathematics 2152, DOI 10.1007/978-3-319-26437-0_6

be the situation we work in when M is a high syzygy over a complete intersection ring. With these assumptions, the module L defined as the dual of the second syzygy of the dual of M is the unique maximal Cohen-Macaulay module without free summands whose second syzygy is M, so both L and $\widetilde{M}$ are determined by M alone; see Lemma 7.1.3.

Remark 6.1.3 Under the assumptions and in the notation of Definition 6.1.1, note that neither M nor L can have free summands since their resolutions have a surjective CI operator.

Remark 6.1.4 Corollary 4.3.4 (which is an immediate corollary of Propositions 4.3.2 and 4.3.3) provides an alternate way of looking at pre-stable syzygies: Under the assumptions and in the notation of Definition 6.1.1, let $(\mathbf{F}, \delta)$ be a minimal R-free resolution of L, and let $(\widetilde{\mathbf{F}}, \tilde{\delta})$ be a lifting of F to a sequence of maps of free modules over $\widetilde{R}$. The minimal free resolution of L as an $\widetilde{R}$-module is the kernel of $\tilde{t} : \widetilde{\mathbf{F}} \longrightarrow \widetilde{\mathbf{F}}[2]$ by Corollary 4.3.4. Thus, we may write the second syzygy $\widetilde{M}$ of L over $\widetilde{R}$ as $\widetilde{M} = \operatorname{Ker}(\tilde{\delta}_1)$, and this module is independent of the choice of the lifting of δ_1 to $\widetilde{R}$.

Stability and pre-stability are preserved under taking syzygies:

Proposition 6.1.5 *Suppose that $f_1, \ldots, f_c$ is a regular sequence in a local ring S, and set $R = S/(f_1, \ldots, f_c)$. If M is a pre-stable syzygy over R, then $\operatorname{Syz}_1^R(M)$ is pre-stable as well. If M is a stable syzygy over R, then so is $\operatorname{Syz}_1^R(M)$.*

Proof Let $(\mathbf{F}, \delta)$ be a minimal R-free resolution of a module L such that $M = \operatorname{Syz}_2^R L$ and the conditions in Definition 6.1.1 are satisfied. Lifting $\mathbf{F}$ to $\widetilde{\mathbf{F}}$ over $\widetilde{R} := S/(f_1, \ldots, f_{c-1})$ and using the hypothesis that S is local, we see that the lifted CI operator $\tilde{t}_c$ is surjective on $\widetilde{F}$. By Propositions 4.3.2 and 4.3.3, $\widetilde{\mathbf{G}} := \operatorname{Ker}(\tilde{t}_c)$ is the minimal free resolution of the module L over $\widetilde{R}$.

Let $M' = \operatorname{Syz}_1^R(M)$ and let $L' = \operatorname{Syz}_1^R(L)$, so that $\mathbf{F}' = \mathbf{F}_{\geq 1}[-1]$ is the minimal free resolution of L'. Clearly $t_c\big|_{\mathbf{F}'}$ is surjective. The shifted truncation $\widetilde{\mathbf{F}}' := \widetilde{\mathbf{F}}_{\geq 1}[-1]$ is a lifting of $\mathbf{F}'$, and $\widetilde{\mathbf{G}}' := \operatorname{Ker}\left(\tilde{t}_c\big|_{\widetilde{\mathbf{F}'}}\right)$ is a minimal free resolution of L' over $\widetilde{R}$. The complex $\widetilde{\mathbf{G}}'_{\geq 2}$ agrees (up to the sign of the differential) with $\widetilde{\mathbf{G}}[-1]_{\geq 2}$:

$$\begin{array}{ll} \widetilde{\mathbf{G}}: & \ldots \longrightarrow \widetilde{G}_4 \longrightarrow \widetilde{G}_3 \longrightarrow \widetilde{G}_2 \longrightarrow \widetilde{F}_1 \xrightarrow{\tilde{\delta}_1} \widetilde{F}_0 \\ \widetilde{\mathbf{G}}': & \ldots \longrightarrow \widetilde{G}_4 \longrightarrow \widetilde{G}_3 \longrightarrow \widetilde{F}_2 \xrightarrow{\tilde{\delta}_2} \widetilde{F}_1 , \end{array} \tag{6.1}$$

Thus

$$\operatorname{Ker}(\tilde{\delta}_2) = \operatorname{Syz}_1^{\widetilde{R}}(\operatorname{Ker}(\tilde{\delta}_1)) .$$

Since $\operatorname{Ker}(\tilde{\delta}_1)$ is a pre-stable syzygy by hypothesis, we can apply the induction hypothesis to conclude that $\operatorname{Ker}(\tilde{\delta}_2)$ is pre-stable.

The last statement in the proposition follows from the observation that if L is a maximal Cohen-Macaulay R-module, and R is a Cohen-Macaulay ring, $L' = \mathrm{Syz}_1^R(L)$ is again a maximal Cohen-Macaulay module. □

The next result shows that in the codimension 1 case, pre-stable syzygies are the same as codimension 1 matrix factorizations.

Proposition 6.1.6 *Let $f \in S$ be a non-zerodivisor in a local ring and set $R = S/(f)$. The following conditions on an R-module M are equivalent:*

(1) *M is a pre-stable syzygy with respect to f.*
(2) *M has projective dimension 1 as an S-module, and has no free summands.*
(3) *The minimal R-free resolution of M comes from a codimension 1 matrix factorization of f over S.*

If S is a Cohen-Macaulay ring, then all these conditions are equivalent to M being a stable syzygy with respect to f.

Proof (1) $\Rightarrow$ (2): Since M is a pre-stable syzygy, it is the second syzygy of an R-module L whose second syzygy as a module over S is pre-stable, and thus 0. It follows that L has projective dimension ≤ 1 as an S-module. As L has no free summands, we conclude by Theorem 2.1.1 that the minimal R-free resolution of L is given by a minimal matrix factorization of f. Since M is a syzygy of L over R, the same is true of M.

(2) $\Rightarrow$ (3): If M has projective dimension 1 then M is the cokernel of a square matrix over S, and the homotopy for multiplication by f defines a matrix factorization.

(3) $\Rightarrow$ (1): Continuing the periodic free resolution of M as an R module two steps to the right we get a minimal free resolution $\mathbf{F}$ of a module $L \cong M$ on which the CI operator is surjective, and also injective on $\mathbf{F}_{\geq 2}$. By Corollary 4.3.4, it follows that the projective dimension of L over S is 1. Hence, $\mathrm{Syz}_2^S(L) = 0$, completing the proof of pre-stability.

If R is a Cohen-Macaulay ring, then we can take $L \cong M$ a maximal Cohen-Macaulay module by (3). □

We now return to the situation of Theorem 4.2.3: Let N be an R-module with finite projective dimension over S. We regard $E := \mathrm{Ext}_R(N, k)$ as a module over $\mathcal{R} = k[\chi_1, \ldots, \chi_c]$, where χ_j have degree 2 and $\mathrm{Ext}_R^i(N, k)$ is in degree i. Since we think of degrees in E as cohomological degrees, we write $E[a]$ for the shifted module whose degree i component is $E^{i+a} = \mathrm{Ext}_R^{i+a}(N, k)$. (Note that this is the shifted module often written as $E(-a)$; since we are using notation such as $M(p)$ for a different purpose, we will stick with the square bracket notation.) If M is the r-th syzygy module of N then $\mathrm{Ext}_R(M, k) = \mathrm{Ext}_R^{\geq r}(N, k)[-r]$.

Recall that the *Castelnuovo-Mumford regularity* $\operatorname{reg} E$ is defined as

$$\operatorname{reg} E = \max_{0 \leq i \leq c} \left\{ i + \{\max\{j \mid H^i_{(\chi_1, \ldots, \chi_c)}(E)^j \neq 0\}\} \right\},$$

where $H^i_{(\chi_1,\dots,\chi_c)}(E)$ denotes the local cohomology module. In particular, we have $\operatorname{reg} E[a] = (\operatorname{reg} E) - a$. Since the generators of $\mathcal{R}$ have degree 2, some care is necessary. Note that if $\operatorname{Ext}_R^{odd}(N,k) \neq 0$ then $E = \operatorname{Ext}_R(N,k)$ can never have regularity ≤ 0, since it is generated in degrees ≥ 0 and the odd part cannot be generated by the even part. Thus we will often have recourse to the condition $\operatorname{reg} \operatorname{Ext}_R(N,k) = 1$. On the other hand, many things work as usual. If we split E into even and odd parts, $E = E^{even} \oplus E^{odd}$ we have

$$\operatorname{reg} E = \max(\operatorname{reg} E^{even}, \operatorname{reg} E^{odd})$$

as usual. Also, if χ_c is a non-zerodivisor on E then $\operatorname{reg}(E/\chi_c E) = \operatorname{reg} E$.

Theorem 6.1.7 *Suppose that $f_1,\dots,f_c$ is a regular sequence in a local ring S with infinite residue field k, and set $R = S/(f_1,\dots,f_c)$. Let N be an R-module with finite projective dimension over S, and let $\mathbf{L}$ be the minimal R-free resolution of N. There exists a non-empty Zariski open dense set $\mathcal{Z}$ of strictly upper-triangular matrices $(\alpha_{i,j})$ with entries in k, such that for every*

$$r \geq 2c - 1 + \operatorname{reg}(\operatorname{Ext}_R(N,k))$$

the syzygy module $\operatorname{Syz}_r^R(N)$ is pre-stable with respect to the regular sequence $f_1',\dots,f_c'$ with

$$f_i' = f_i + \sum_{j>i} \alpha_{i,j} f_j \,.$$

To prepare for the proof of Theorem 6.1.7 we will explain the property of the regular sequence $f_1',\dots,f_c'$ that we will use. Recall that a sequence of elements $\chi_c', \chi_{c-1}',\dots,\chi_1' \in \mathcal{R}$ is said to be an *almost regular sequence* on a graded module E if, for $q = c,\dots,1$, the submodule of elements of $E/(\chi_{q+1}',\dots,\chi_c')E$ annihilated by χ_q' is of finite length.

We will use the following lemma with $E = \operatorname{Ext}_R(N,k)$. As before, we denote by $E[a]$ the shifted module with $E[a]^i = E^{a+i}$.

Lemma 6.1.8 *Suppose that $E = \oplus_{i\geq 0} E^i$ is a graded module of regularity ≤ 1 over $\mathcal{R} = k[\chi_1,\dots,\chi_c]$. The element χ_c is almost regular on E if and only if χ_c is a non-zerodivisor on $E^{\geq 2}[2]$ (equivalently, χ_c is a non-zerodivisor on $E^{\geq 2}$).*

More generally, if we set $E(c) = E$ and

$$E(j-1) = E(j)^{\geq 2}[2]/\chi_j E(j)^{\geq 2}[2]$$

for $j \leq c$, then the sequence $\chi_c,\dots,\chi_1$ is almost regular on E if and only if χ_j is a non-zerodivisor on $E(j)^{\geq 2}[2]$ for every j. In that case $\operatorname{reg} E(i) \leq 1$.

Proof By definition the element χ_c is almost regular on E if the submodule P of E of elements annihilated by χ_c has finite length. Since $\text{reg}(E) \leq 1$, all such elements must be contained in $E^{\leq 1}$. Hence, χ_c is a non-zerodivisor on $E^{\geq 2}$.

Conversely, if χ_c is a non-zerodivisor on $E^{\geq 2}$ then $P \subseteq E^{\leq 1}$ so P has finite length. Therefore, χ_c is almost regular on E.

Thus χ_c is almost regular if and only if it is a non-zerodivisor on $E^{\geq 2}$ as claimed. If χ_c is a non-zerodivisor on $E^{\geq 2}$, then

$$\text{reg}(E^{\geq 2}/\chi_c E^{\geq 2}) = \text{reg}(E^{\geq 2}) \leq 3,$$

whence $\text{reg}(E(c-1)) \leq 1$. By induction, $\chi_{c-1}, \dots, \chi_1$ is an almost regular sequence on $E(c-1)$ if and only if χ_j is a non-zerodivisor on $E(j)^{\geq 2}[2]$ for every $j < c$, as claimed. □

The following result is a well-known consequence of the "Prime Avoidance Lemma" (see for example [26, Lemma 3.3] for Prime Avoidance):

Lemma 6.1.9 *If k is an infinite field and E is a graded non-zero module over the polynomial ring $\mathcal{R} = k[\chi_1, \dots, \chi_c]$ of regularity ≤ 1, then there exists a non-empty Zariski open dense set $\mathcal{Y}$ of lower-triangular matrices $(v_{i,j})$ with entries in k, such that the sequence of elements $\chi'_c, \dots, \chi'_1$ with*

$$\chi'_i = \chi_i + \sum_{j<i} v_{i,j}\chi_j$$

is almost regular on E.

For the reader's convenience we give a short proof.

Proof The proof is by induction on c. Since $\text{reg}(E) \leq 1$, the maximal submodule of E of finite length is contained in $E^{\leq 1}$. Therefore, $\mathbf{m} = (\chi_1, \dots, \chi_c)$ is not an associated prime of $E' = E^{\geq 2}[2]$. Let $P_1, \dots, P_u$ be the associated primes of E'. None of them can contain the entire set

$$\left\{\chi_c + \sum_{1\leq m\leq c-1} a_m\chi_m \,\middle|\, a_m \in k\right\}$$

since that set generates $\mathbf{m}$. As the field k is infinite, by the Prime Avoidance Lemma (see for example [26, Lemma 3.3] for Prime Avoidance) it follows that there exists a non-empty Zariski open dense set of vectors $(a_1, \dots, a_{c-1})$ such that

$$\chi'_c = \chi_c + \sum_{1\leq m\leq c-1} a_m\chi_m$$

is not in the set of zerodivisors $P_1 \cup \cdots \cup P_u$. Thus, χ'_c is a non-zerodivisor on E' as desired.

Since $\text{reg}(E^{\geq 2}) \leq 3$, we conclude that $\text{reg}(E'/\chi'_c E') \leq 1$. □

Remark 6.1.10 Again let $f_1, \dots, f_c$ be a regular sequence in a local ring S with infinite residue field k and maximal ideal $\mathbf{m}$, and set $R = S/(f_1, \dots, f_c)$. Let N be an R-module with finite projective dimension over S, and let $\mathbf{L}$ be the minimal R-free resolution of N. Suppose we have CI operators defined by a lifting $\widetilde{\mathbf{L}}$. If we make a change of generators of $(f_1, \dots, f_c)$ using an invertible matrix α and $f_i' = \sum_j \alpha_{i,j} f_j$ with $\alpha_{i,j} \in S$, then the lifted CI operators on the lifting $\widetilde{\mathbf{L}}$ change as follows:

$$\tilde{\partial}^2 = \sum_i f_i' \tilde{t}_i' = \sum_i \left(\sum_j \alpha_{i,j} f_j \right) \tilde{t}_i' = \sum_j f_j \left(\sum_i \alpha_{i,j} \tilde{t}_i' \right).$$

So the CI operators corresponding to the sequence $f_1, \dots, f_c$ are expressed as $t_j = \sum_i \alpha_{i,j} t_i'$. Thus, if we make a change of generators of the ideal $(f_1, \dots, f_c)$ using a matrix α then the CI operators transform by the inverse of the transpose of α.

In view of this remark, Lemmas 6.1.8 and 6.1.9 can be translated as follows:

Proposition 6.1.11 *Let $f_1, \dots, f_c \in S$ be a regular sequence in a local ring with infinite residue field k, and set $R := S/(f_1, \dots, f_c)$. Let N be an R-module of finite projective dimension over S, and set $E := \mathrm{Ext}_R(N, k)$.*

(1) [4, 25] *There exists a non-empty Zariski open dense set $\mathcal{Z}$ of upper-triangular matrices $\overline{\alpha} = (\overline{\alpha}_{i,j})$ with entries in k, such that if $\alpha = (\alpha_{i,j})$ is any matrix over S that reduces to $\overline{\alpha}$ modulo the maximal ideal of S, and $\nu = (\alpha^\vee)^{-1}$, then the sequence $f_1', \dots, f_c'$ with*

$$f_i' = f_i + \sum_{j>i} \alpha_{i,j} f_j$$

corresponds to a sequence of CI operators $\chi_c', \dots, \chi_1'$, where $\chi_i' = \sum_j \nu_{i,j} \chi_j$, that is almost regular on E.

(2) *Set $E(c) = E$ and*

$$E(i-1) = E(i)^{\geq 2}[2] / \chi_i' E(i)^{\geq 2}[2]$$

for $i \leq c$, and suppose that $\mathrm{reg}(E) \leq 1$. *If $\nu = (\alpha^\vee)^{-1}$. If χ_i' is defined as in* (1), *then χ_i' is a non-zerodivisor on $E(i)^{\geq 2}[2]$ for every i.* □

We say that $f_1', \dots, f_c'$ with

$$f_i' = f_i + \sum_{j>i} \alpha_{i,j} f_j$$

are *generic* for N if $(\alpha_{i,j}) \in \mathcal{Z}$ in the sense above.

Proof of Theorem 6.1.7 To simplify the notation, we may begin by replacing N by its $(\mathrm{reg}(\mathrm{Ext}_R(N,k))-1)$-st syzygy, and assume that $\mathrm{reg}(\mathrm{Ext}_R(N,k)) = 1$. After a general change of $f_1,\dots,f_c$ we may also assume, by Lemma 6.1.8, that $\chi_c,\dots,\chi_1$ is an almost regular sequence on $\mathrm{Ext}_R(N,k)$. By Proposition 6.1.5 it suffices to treat the case $r=2c$. Set $M=\mathrm{Syz}^R_{2c}(N)$.

Let $(\mathbf{F},\delta)$ be the minimal free resolution of $N' := \mathrm{Syz}^R_2(N)$, so that $M = \mathrm{Ker}(\delta_{2c-3})$. Since N has finite projective dimension over S, the module N' also has finite projective dimension over S.

Let $(\widetilde{\mathbf{F}},\tilde\delta)$ be a lifting of $\mathbf{F}$ to $\widetilde R := S/(f_1,\dots,f_{c-1})$, and let $\tilde t_c$ be the lifted CI operator. Set $(\widetilde{\mathbf{G}},\tilde\delta) = \mathrm{Ker}(\tilde t_c)$. By Proposition 6.1.11, χ_c is a monomorphism on $\mathrm{Ext}_R(N',k) = \mathrm{Ext}^{\geq 2}_R(N,k)[2]$. Since χ_c is induced by t_c, Nakayama's Lemma implies that t_c is surjective, so in particular $\mathbf{F}_{\geq 2c-2} \longrightarrow \mathbf{F}_{\geq 2c-4}$ is surjective, as required for Condition (1) in 6.1.1 for $c>1$.

Using Nakayama's Lemma again, we see that the lifted CI operator $\tilde t_c$ is also an epimorphism. Propositions 4.3.2 and 4.3.3 show that $\widetilde{\mathbf{G}}$ is a minimal free resolution of N' over $\widetilde R$, and $\mathbf{F}$ is obtained from $\widetilde{\mathbf{G}}$ by the Shamash construction 4.1.3. Hence

$$\mathrm{Ext}_{\widetilde R}(N',k) = \mathrm{Ext}_R(N',k)/\chi_c\mathrm{Ext}_R(N',k)\,,$$

and therefore

$$\mathrm{Ext}_{\widetilde R}(N',k) = \Big(\mathrm{Ext}^{\geq 2}_{\widetilde R}(N,k)\Big/\chi_c\mathrm{Ext}^{\geq 2}_{\widetilde R}(N,k)\Big)[2].$$

By Proposition 6.1.11, χ_c is a non-zerodivisor on $\mathrm{Ext}^{\geq 2}_{\widetilde R}(N,k)$, and we conclude that $\mathrm{Ext}_{\widetilde R}(N',k)$ has regularity ≤ 1 over $k[\chi_1,\dots,\chi_{c-1}]$.

We will prove the theorem by induction on c.

Suppose $c=1$. Then $M=N'$ is the second syzygy of N. In this case $\widetilde R = S$, and by hypothesis $M=N'$ has finite projective dimension over S. Therefore, $\mathrm{Ext}_S(M,k)$ is a module of finite length. Since it has regularity ≤ 1 (as a module over k), it follows that it is zero except in degrees ≤ 1, that is, the projective dimension of M over $\widetilde R$ is ≤ 1. By Proposition 6.1.6, M is a pre-stable syzygy.

Now suppose that $c>1$. Note that $M=\mathrm{Syz}^R_{2c-2}(N')$, and following the notation in Definition 6.1.1 set $L=\mathrm{Syz}^R_{2c-4}(N')$. Then $M=\mathrm{Syz}^R_2(L)$. We consider the module $\widetilde M := \mathrm{Syz}^{\widetilde R}_2(L)$. Propositions 4.3.2 and 4.3.3 show that

$$\widetilde M = \mathrm{Ker}(\tilde\delta_{2c-3}) = \mathrm{Syz}^{\widetilde R}_{2(c-1)}(N')\,.$$

By induction hypothesis, $\widetilde M$ is a pre-stable syzygy, verifying Condition (2) in 6.1.1. Thus M is a pre-stable syzygy. □

The following lemma is well-known:

Lemma 6.1.12 *Let R be a local Cohen-Macaulay ring. If N is an R-module, then the depth of* $\mathrm{Syz}_i(N)$ *increases strictly with i until it reaches the maximal possible value* $\mathrm{depth}(R)$, *and after that it stays constant.*

Proof Let **F** be the minimal free resolution of N. Compute depth using the short exact sequences

$$0 \longrightarrow \mathrm{Syz}_{i+1}(N) \longrightarrow F_i \longrightarrow \mathrm{Syz}_i(N) \longrightarrow 0\,.$$

□

Combining Lemma 6.1.12 with the proof of Theorem 6.1.7, we will show that sufficiently high syzygies over complete intersections are indeed stable:

Proof of Theorem 6.1.2 We replace the module N in Theorem 6.1.7 by a maximal Cohen-Macaulay syzygy; Lemma 6.1.12 shows that this can be done.

We will analyze the proof of Theorem 6.1.7, and use its notation. Note that N has $\mathrm{depth}_R(N) \geq \mathrm{depth}(R) - 1$. By Lemma 6.1.12, all syzygies of N are maximal Cohen-Macaulay modules. In particular, the module L constructed in the proof of Theorem 6.1.7 is a syzygy of N, and thus is a maximal Cohen-Macaulay module. Furthermore, the proof by induction of Theorem 6.1.7 works out since the module N', constructed there, is a syzygy of N, and so $\mathrm{depth}_{\tilde{R}}(N') \geq \mathrm{depth}(\tilde{R}) - 1$ is satisfied. □

6.2 The Graded Case

Chapters 3 and 5 give constructions producing graded resolutions (see [48]) if we start with a graded higher matrix factorization.

We can use the theory above to obtain graded higher matrix factorizations in the graded case as long as the given regular sequence $f_1, \ldots, f_c$ consists of elements of the same degree, and thus a general linear scalar combination of them is still homogeneous. In this case, Proposition 6.1.11 and Theorem 6.1.7 hold for $E = \mathrm{Ext}_R(N, k)$ verbatim, without first localizing at the maximal ideal. We can analyze the two-variable Poincaré series of a stable syzygy module:

Corollary 6.2.1 *Let k be an infinite field, $S = k[x_1, \ldots, x_n]$ be standard graded with $\deg(x_i) = 1$ for each i, and I be an ideal generated by a regular sequence of c homogeneous elements of the same degree q. Set $R = S/I$, and suppose that N is a finitely generated graded R-module. Let $f_1, \ldots, f_c$ be a generic for N regular sequence of forms minimally generating I. If M is a sufficiently high graded syzygy of N over R, then M is the module of a minimal higher matrix factorization (d, h) with respect to $f_1, \ldots, f_c$; it involves modules $B_s(p)$ for $s = 0, 1$ and $1 \leq p \leq c$. Denote*

$$\beta^R_{i,j}(M) = \dim\left(\mathrm{Tor}^R_i(M, k)_j\right)$$

the graded Betti numbers of M over R. The graded Poincaré series

$$\mathcal{P}^R_M(x, z) = \sum_{i \geq 0} \beta^R_{i,j}(M) x^i z^j$$

of M over R is

$$\mathcal{P}_M^R(x,z) = \sum_{1\le p\le c} \frac{1}{(1-x^2z^q)^{c-p+1}} \Big(x\, m_{p;1}(z) + m_{p;0}(z) \Big)\,, \tag{6.2}$$

where

$$m_{p;s}(z) := \sum_{j\ge 0} \beta^S_{s,j}\Big(B_s(p)\Big) z^j$$

in which $\beta^S_{s,j}\Big(B_s(p)\Big)$ *denotes the number of minimal generators of degree* j *of the* S*-free module* $B_s(p)$ *(and as usual,* $1 \le p \le c,\ s = 0, 1$*).*

Proof Note that (6.2) is a refined version of the formula in Corollary 5.2.1(1). The CI operators t_i on the minimal R-free resolution **T**, constructed in 5.1.1, of the HMF module M can be taken homogeneous. Writing $\tilde{t}_i$ for a lift to S of the CI operator t_i we have

$$\tilde{d}^2 = f_1\tilde{t}_1 + \cdots + f_c\tilde{t}_c$$

and $\deg(t_i) = -q$ for every i. □

In [29] we will use the formulas in Corollary 6.2.1 to study quadratic complete intersections.

If the forms $f_1, \ldots, f_c$ do not have the same degree there may be no "sufficiently general choice" of generators for $(f_1, \ldots, f_c)$ consisting of homogeneous elements, as the following example shows:

Example 6.2.2 Let

$$R = k[x, y]/(x^2, y^3)$$

and consider the module

$$N = R/x \oplus R/y\,.$$

Over the local ring $S_{(x,y)}/(x^2, y^3)$ the CI operator corresponding to $x^2 + y^3$ is eventually surjective. However, the minimal R-free resolution of N is the direct sum of the free resolutions of R/x and R/y. The CI operator corresponding to x^2 vanishes on the minimal free resolution of R/y. The CI operator corresponding to y^3 vanishes on the minimal free resolution of R/x, and hence the CI operator corresponding to $y^3 + ax^3 + bx^2y$, for any a, b, does too. Thus, there is no homogeneous linear combination of x^2, y^3 that corresponds to an eventually surjective CI operator.

6.3 The Box Complex

Suppose that $f \in S$ is a non-zerodivisor. Given an S-free resolution of an $S/(f)$-module L and a homotopy for f, we will construct an S-free resolution of the second syzygy $\mathrm{Syz}_2^{S/(f)}(L)$ of L as an $S/(f)$-module, and also a homotopy for f on it, using a mapping cone in a special circumstance, which we call the Box complex.

Box Construction 6.3.1 Suppose that $f \in S$ and

$$\mathbf{Y}: \quad \cdots \longrightarrow Y_4 \xrightarrow{\partial_4} Y_3 \xrightarrow{\partial_3} Y_2 \underset{\theta_1}{\overset{\partial_2}{\rightleftarrows}} Y_1 \underset{\theta_0}{\overset{\partial_1}{\rightleftarrows}} Y_0$$

is a free complex over S with homotopies $\{\theta_i : Y_i \longrightarrow Y_{i+1}\}_{i=0,1}$ for f. We call the mapping cone

$$\mathrm{Box}(\mathbf{Y}): \quad \cdots \longrightarrow Y_4 \xrightarrow{\partial_4} \begin{matrix} Y_3 \\ \oplus \\ Y_1 \end{matrix} \xrightarrow{\begin{pmatrix} \partial_3 & \psi \\ 0 & \partial_1 \end{pmatrix}} \begin{matrix} Y_2 \\ \oplus \\ Y_0 \end{matrix} \qquad (6.3)$$

of the map

$$\psi := \theta_1 : \mathbf{Y}_{\leq 1}[1] \longrightarrow \mathbf{Y}_{\geq 2}$$

the *box complex* and denote it $\mathrm{Box}(\mathbf{Y})$.

Box Theorem 6.3.2 *Under the assumptions and with notation as in the Box Construction* 6.3.1, *if f is a non-zerodivisor on S and* $\mathbf{Y}$ *is an S-free resolution of a module L annihilated by f, then* $\mathrm{Box}(\mathbf{Y})$ *is an S-free resolution of the second $S/(f)$-syzygy of L.*

Moreover, with notation as in diagram (6.4)

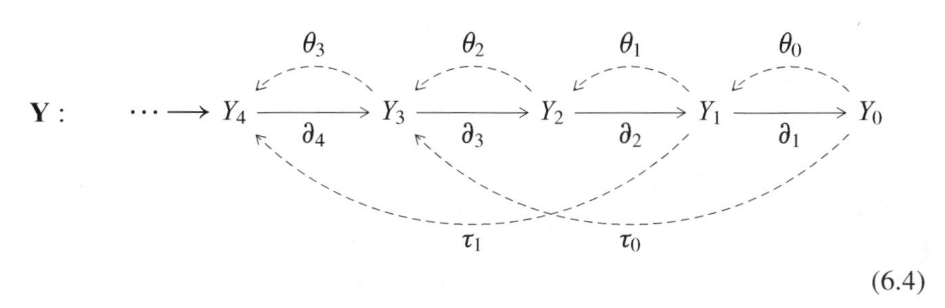

(6.4)

suppose that $\{\theta_i : Y_i \longrightarrow Y_{i+1}\}_{i\geq 0}$ *is a homotopy for* f *on* $\mathbf{Y}$ *and*

$$\tau_0 : Y_0 \longrightarrow Y_3$$
$$\tau_1 : Y_1 \longrightarrow Y_4$$

are higher homotopies for f *(which exist by Proposition* 3.4.2*), so that*

$$\partial_3\tau_0 + \theta_1\theta_0 = 0$$
$$\tau_0\partial_1 + \theta_2\theta_1 + \partial_4\tau_1 = 0\,.$$

Then the maps

$$\begin{pmatrix}\theta_2 & \tau_0\\ \partial_2 & \theta_0\end{pmatrix},\ (\theta_3, \tau_1),\ \theta_4,\ \theta_5,\ \ldots \tag{6.5}$$

give a homotopy for f *on* $\mathrm{Box}(\mathbf{Y})$ *as shown in diagram* (6.6)*:*

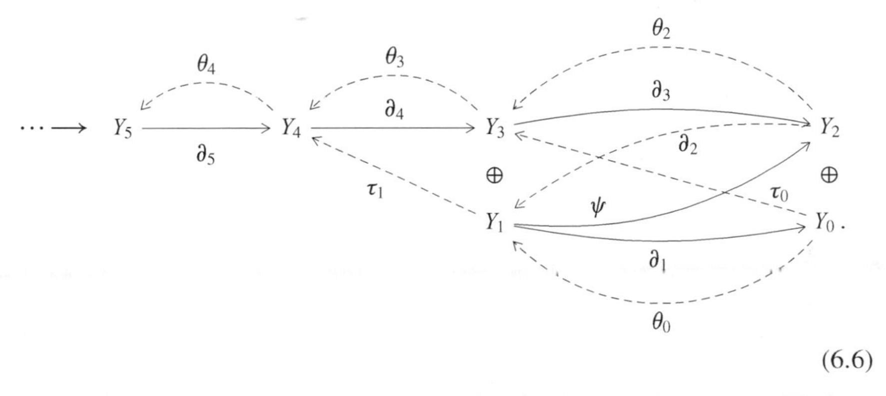

(6.6)

A similar formula yields a full system of higher homotopies on $\mathrm{Box}(\mathbf{Y})$ from higher homotopies on $\mathbf{Y}$, but we will not need this.

Proof The following straightforward computation shows that the maps in (6.5) are homotopies for f on $\mathrm{Box}(\mathbf{Y})$:

$$\begin{pmatrix}\partial_3 & \theta_1\\ 0 & \partial_1\end{pmatrix}\begin{pmatrix}\theta_2 & \tau_0\\ \partial_2 & \theta_0\end{pmatrix} = \begin{pmatrix}\partial_3\theta_2 + \theta_1\partial_2 & \partial_3\tau_0 + \theta_1\theta_0\\ \partial_1\partial_2 & \partial_1\theta_0\end{pmatrix} = \begin{pmatrix}f & 0\\ 0 & f\end{pmatrix} \tag{6.7}$$

$$\begin{pmatrix}\partial_1 & \theta_1\\ 0 & \partial_1\end{pmatrix} + \begin{pmatrix}\partial_4\theta_3 & \partial_4\tau_1\\ 0 & 0\end{pmatrix} = \begin{pmatrix}\theta_2\partial_3 + \partial_4\theta_3 & \theta_2\theta_1 + \tau_0\partial_1 + \partial_4\tau_1\\ \partial_2\partial_3 & \partial_2\theta_1 + \theta_0\partial_1\end{pmatrix} = \begin{pmatrix}f & 0\\ 0 & f\end{pmatrix}.$$

Next we will prove that $\mathrm{Box}(\mathbf{Y})$ is a resolution. Consider the short exact sequence of complexes

$$0 \longrightarrow \mathbf{Y}_{\geq 2} \longrightarrow \mathrm{Box}(\mathbf{Y}) \longrightarrow \mathbf{Y}_{\leq 1} \longrightarrow 0\,.$$

Since $\mathbf{Y}_{\leq 1}$ is a two-term complex,

$$H_i(\mathrm{Box}(\mathbf{Y})) = H_i(\mathbf{Y}_{\geq 2}) = 0$$

for $i \geq 2$. If $(v, w) \in Y_3 \oplus Y_1$ is a cycle in $\mathrm{Box}(\mathbf{Y})$, then applying the homotopy maps in (6.5) we get:

$$(fv, fw) = (\partial_4\theta_3(v) + \partial_4\tau_1(w), 0)\,.$$

Since f is a non-zerodivisor, it follows that $w = 0$ and thus v is a cycle in $\mathbf{Y}_{\geq 2}$. Since $\mathbf{Y}_{\geq 2}$ is acyclic, v is a boundary in $\mathbf{Y}_{\geq 2}$. Hence, the complex $\mathrm{Box}(\mathbf{Y})$ is acyclic.

To simplify notation, we write $\overline{-}$ for $S/(f) \otimes -$ and set $\psi = \theta_1$. To complete the proof we will show that

$$H_0(\mathrm{Box}(\mathbf{Y})) = \mathrm{Ker}(\overline{\partial_1} : \overline{Y}_1 \longrightarrow \overline{Y}_0)\,.$$

Since we have a homotopy for f on $\mathbf{Y}$, we see that f annihilates the module resolved by $\mathbf{Y}$. Therefore, $H_0(\mathbf{Y}) = H_1(\overline{\mathbf{Y}})$. The complex $\overline{\mathrm{Box}(\mathbf{Y})}$ is the mapping cone $\mathbf{Cone}(\overline{\psi} \otimes S/(f))$, where $\overline{\psi} = \psi \otimes S/(f)$, so there is an exact sequence of complexes

$$0 \longrightarrow \overline{\mathbf{Y}}_{\geq 2} \longrightarrow \overline{\mathrm{Box}(\mathbf{Y})} \longrightarrow \overline{\mathbf{Y}}_{\leq 1} \longrightarrow 0\,.$$

Since $\mathbf{Y}$ is a resolution, $H_0(\mathbf{Y}_{\geq 2})$ is contained in the free S-module Y_1. Thus f is a non-zerodivisor on $H_0(\mathbf{Y}_{\geq 2})$ and $\overline{\mathbf{Y}}_{\geq 2}$ is acyclic. The long exact sequence for the mapping cone now yields

$$0 \longrightarrow H_1(\mathbf{Cone}(\overline{\psi})) \longrightarrow H_1(\overline{\mathbf{Y}}_{\leq 1}) \xrightarrow{\overline{\psi}} H_0(\overline{\mathbf{Y}}_{\geq 2})\,.$$

It suffices to prove that the map induced on homology by $\overline{\psi}$ is 0. Let $u \in Y_1$ be such that $\overline{u} \in \mathrm{Ker}(\overline{\partial}_1)$, so $\partial_1(u) = fy$ for some $y \in Y_0$. We also have $fy = \partial_1\theta_0(y)$, so $u - \theta_0(y) \in \mathrm{Ker}(\partial_1)$. Since $\mathbf{Y}$ is acyclic $u = \theta_0(y) + \partial_2(z)$ for some $z \in Y_2$. Applying ψ we get

$$\begin{aligned}
\psi(u) &= \theta_1\theta_0(y) + \theta_1\partial_2(z) \\
&= -\partial_3\tau_0(y) + \Big(fz - \partial_3\theta_2(z)\Big) \\
&= -\partial_3\Big(\tau_0(y) + \theta_2(z)\Big) + fz,
\end{aligned}$$

so the map induced on homology by $\overline{\psi}$ is 0 as desired. □

Theorem 6.3.2 has a partial converse that we will use in the proof of Theorem 7.2.1.

Proposition 6.3.3 *Let $f \in S$ be a non-zerodivisor and set $R = S/(f)$. Let*

$$\begin{array}{ccccc} \cdots \to Y_4 & \xrightarrow{\partial_4} & Y_3 & \xrightarrow{\partial_3} & Y_2 \\ & & \oplus & \psi \nearrow & \oplus \\ & & Y_1 & \xrightarrow{\partial_1} & Y_0 \end{array}$$

be an S-free resolution of a module annihilated by f. Set $\theta_1 := \psi$, and with notation as in diagram (6.6), suppose that

$$\begin{pmatrix} \theta_2 & \tau_0 \\ \partial_2 & \theta_0 \end{pmatrix}, \ (\theta_3, \tau_1), \ \theta_4, \ \theta_5, \ \ldots$$

is a homotopy for f. If the cokernels of ∂_2 and of ∂_3 are f-torsion free, then the complex

$$\mathbf{Y}: \quad \ldots \longrightarrow Y_4 \xrightarrow{\partial_4} Y_3 \xrightarrow{\partial_3} Y_2 \xrightarrow{\partial_2} Y_1 \xrightarrow{\partial_1} Y_0 \tag{6.8}$$

is exact and it has homotopies for f as in (6.4).

Proof We first show that $\mathbf{Y}$ is a complex. The equation $\partial_3\partial_4 = 0$ follows from our hypothesis. To show that $\partial_2\partial_3 = 0$ and $\partial_1\partial_2 = 0$, use the homotopy equations

$$\begin{aligned} 0\theta_3 + \partial_2\partial_3 &= 0: \ Y_3 \longrightarrow Y_1 \\ \partial_1\partial_2 &= 0: \ Y_2 \longrightarrow Y_0 \,. \end{aligned}$$

The equalities in (6.7) imply that

$$\begin{aligned} \theta_0 &: \ Y_0 \longrightarrow Y_1 \\ \psi = \theta_1 &: \ Y_1 \longrightarrow Y_2 \\ \theta_2 &: \ Y_2 \longrightarrow Y_3 \\ \theta_3 &: \ Y_3 \longrightarrow Y_4 \end{aligned}$$

form the beginning of a homotopy for f on $\mathbf{Y}$. Thus the complex $\mathbf{Y}$ becomes exact after inverting f. The exactness of $\mathbf{Y}$ is equivalent to the statement that the induced maps $\mathrm{Coker}(\partial_3) \longrightarrow Y_1$ and $\mathrm{Coker}(\partial_2) \longrightarrow Y_2$ are monomorphisms. Since this is true after inverting f, and since the cokernels are f-torsion free by hypothesis, exactness holds before inverting f as well.

Furthermore, (6.7) imply that we have the desired homotopies on $\mathbf{Y}$. □

6.4 From Syzygies to Higher Matrix Factorizations

Higher matrix factorizations arising from pre-stable syzygies have an additional property. We introduce the concept of a pre-stable matrix factorization to capture it.

Definition 6.4.1 A higher matrix factorization (d, h) is a *pre-stable matrix factorization* if, in the notation of 1.4.1, for each $p = 1, \dots, c$ the element f_p is a non-zerodivisor on the cokernel of the composite map

$$R(p-1) \otimes A_0(p-1) \hookrightarrow R(p-1) \otimes A_0(p)$$
$$\xrightarrow{h_p} R(p-1) \otimes A_1(p) \xrightarrow{\pi_p} R(p-1) \otimes B_1(p).$$

If S is Cohen-Macaulay and if the cokernel of the composite map above is a maximal Cohen-Macaulay $R(p-1)$-module, then we say that the higher matrix factorization (d, h) is a *stable matrix factorization*.

We actually do not know of pre-stable matrix factorizations that are not stable. The advantage of stable matrix factorizations over pre-stable matrix factorizations is that if $g \in S$ is an element such that $g, f_1, \dots, f_c$ is a regular sequence and (d, h) is a stable matrix factorization, then $\left(S/(g) \otimes d,\ S/(g) \otimes h\right)$ is again a stable matrix factorization.

Theorem 6.4.2 *Suppose that $f_1, \dots, f_c$ is a regular sequence in a local ring S, and set $R = S/(f_1, \dots, f_c)$. If M is a pre-stable syzygy over R with respect to $f_1, \dots, f_c$, then M is the HMF module of a minimal pre-stable matrix factorization (d, h) such that d and h are liftings to S of the first two differentials in the minimal R-free resolution of M. If M is a stable syzygy, then (d, h) is stable as well.*

Combining Theorems 6.4.2 and 6.1.7 we obtain the following more precise version of Theorem 1.3.1 in the introduction.

Corollary 6.4.3 *Suppose that $f_1, \dots, f_c$ is a regular sequence in a local ring S with infinite residue field k, and set $R = S/(f_1, \dots, f_c)$. Let N be an R-module with finite projective dimension over S. There exists a non-empty Zariski open dense set $\mathcal{Z}$ of matrices $(\alpha_{i,j})$ with entries in k such that for every*

$$r \geq 2c - 1 + \operatorname{reg}(\operatorname{Ext}_R(N, k))$$

the syzygy $\operatorname{Syz}_r^R(N)$ is the module of a minimal pre-stable matrix factorization with respect to the regular sequence $\{f_i' = \sum_j \alpha_{i,j} f_j\}$. □

Proof of Theorem 6.4.2 The proof is by induction on c. If $c = 0$, then $M = 0$ so we are done.

Suppose $c \geq 1$. We use the notation of Definition 6.1.1. By assumption, the CI operator t_c is surjective on a minimal R-free resolution $(\mathbf{F}, \delta)$ of a module L of which

M is the second syzygy. Let $(\widetilde{\mathbf{F}}, \tilde{\delta})$ be a lifting of $(\mathbf{F}, \delta)$ to $R' = S/(f_1, \dots, f_{c-1})$. Since S is local, the lifted CI operator $\tilde{t}_c := (1/f_c)\tilde{\delta}^2$ is also surjective, and we set $(\widetilde{\mathbf{G}}, \tilde{\partial}) := \mathrm{Ker}(\tilde{t}_c)$. By Proposition 4.3.3, $\mathbf{F}$ is the result of applying the Shamash construction to $\widetilde{\mathbf{G}}$. Let $\widetilde{B}_1(c)$ and $\widetilde{B}_0(c)$ be the liftings to R' of F_1 and F_0 respectively. By Propositions 4.3.2 and 4.3.3 the minimal R'-free resolution of L has the form

$$\begin{aligned} \dots \longrightarrow \widetilde{G}_4 \xrightarrow{\tilde{\partial}_4} \widetilde{A}_1(c-1) := \widetilde{G}_3 \\ \xrightarrow{\tilde{\partial}_3} \widetilde{A}_0(c-1) := \widetilde{G}_2 \xrightarrow{\tilde{\partial}_2} \widetilde{B}_1(c) \xrightarrow{\tilde{b}} \widetilde{B}_0(c)\,, \end{aligned} \tag{6.9}$$

where $\tilde{b} := \tilde{\partial}_1$, $\tilde{\partial}_2$, $\tilde{\partial}_3$, $\tilde{\partial}_4$ are the liftings of the differential in $\mathbf{F}$.

Since L is annihilated by f_c there exist homotopy maps $\tilde{\theta}_0, \tilde{\psi} := \tilde{\theta}_1, \tilde{\theta}_2$ and a higher homotopy $\tilde{\tau}_0$ so that on

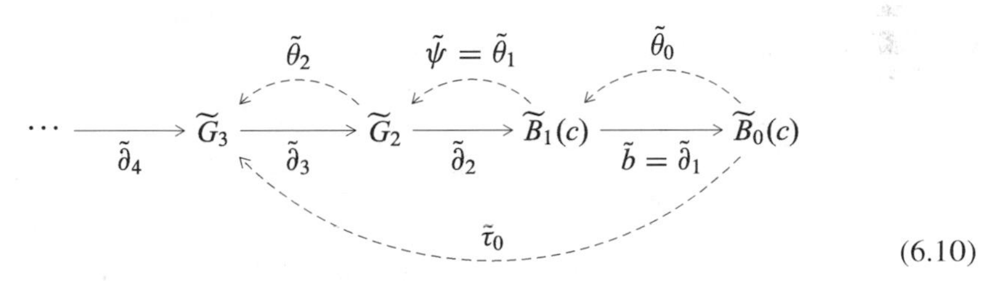

(6.10)

we have:

$$\begin{aligned} \tilde{\partial}_1\tilde{\theta}_0 &= f_c\mathrm{Id} \\ \tilde{\partial}_2\tilde{\theta}_1 + \tilde{\theta}_0\tilde{\partial}_1 &= f_c\mathrm{Id} \\ \tilde{\partial}_3\tilde{\theta}_2 + \tilde{\theta}_1\tilde{\partial}_2 &= f_c\mathrm{Id} \\ \tilde{\partial}_3\tilde{\tau}_0 + \tilde{\theta}_1\tilde{\theta}_0 &= 0\,. \end{aligned} \tag{6.11}$$

Theorem 6.3.2 implies that the minimal free resolution of M over R' has the form:

$$\begin{array}{ccccc} \dots \longrightarrow \widetilde{G}_4 & \xrightarrow{\tilde{\partial}_4} & \widetilde{G}_3 & \xrightarrow{\tilde{\partial}_3} & \widetilde{G}_2 \\ & & \oplus & \tilde{\psi}\nearrow & \oplus \\ & & \widetilde{B}_1(c) & \xrightarrow{\tilde{b}} & \widetilde{B}_0(c) \end{array} \tag{6.12}$$

Using this structure we change the lifting of the differential δ_3 so that

$$\tilde{\delta}_3 = \begin{pmatrix} \tilde{\partial}_3 & \tilde{\psi} \\ 0 & \tilde{b} \end{pmatrix}.$$

Note that the differential $\tilde{\partial}$ on $\widetilde{\mathbf{G}}_{\geq 2}$ has not changed.

Set $M' = \mathrm{Coker}(\widetilde{\mathbf{G}}_{\geq 2}) = \mathrm{Syz}_2^{R'}(L)$. Since M' is a pre-stable syzygy, the induction hypothesis implies that M' is the HMF module of a higher matrix factorization (d', h') with respect to $f_1, \dots, f_{c-1}$ so that the differential $\widetilde{G}_3 \longrightarrow \widetilde{G}_2$ is $\tilde{\partial}_3 = d' \otimes R'$ and the differential $\widetilde{G}_4 \longrightarrow \widetilde{G}_3$ is $\tilde{\partial}_4 = h' \otimes R'$. Thus, there exist free S-modules $A_1'(c-1)$ and $A_0'(c-1)$ with filtrations so that

$$\widetilde{G}_3 = A_1'(c-1) \otimes R' \quad \text{and} \quad \widetilde{G}_2 = A_0'(c-1) \otimes R' \,.$$

We can now define a higher matrix factorization for M. Let $B_1(c)$ and $B_0(c)$ be free S-modules such that $\widetilde{B}_0(c) = B_0(c) \otimes R'$ and $\widetilde{B}_1(c) = B_1(c) \otimes R'$. For $s = 0, 1$, we consider free S-modules A_1 and A_0 with filtrations such that $A_s(p) = A_s'(p)$ for $1 \leq p \leq c-1$ and

$$A_s(c) = A_s'(c-1) \oplus B_s(c) \,.$$

We define the map $d : A_1 \longrightarrow A_0$ to be

$$A_1(c) = A_1(c-1) \oplus B_1(c) \xrightarrow{\begin{pmatrix} d' & \psi_c \\ 0 & b_c \end{pmatrix}} A_0(c-1) \oplus B_0(c) = A_0(c) \qquad (6.13)$$

where b_c and ψ_c are arbitrary lifts to S of $\tilde{b}$ and $\tilde{\psi}$. For every $1 \leq p \leq c-1$, we set $h_p = h_p'$. Furthermore, we define

$$h_c : A_0(c) = A_0 \longrightarrow A_1(c) = A_1$$

to be

$$A_0(c) = A_0(c-1) \oplus B_0(c) \xrightarrow{\begin{pmatrix} \theta_2 & \tau_0 \\ \partial_2 & \theta_0 \end{pmatrix}} A_1(c-1) \oplus B_1(c) = A_1(c) \qquad (6.14)$$

where $\theta_2, \partial_2, \theta_0, \tau_0$ are arbitrary lifts to S of $\tilde{\theta}_2, \tilde{\partial}_2, \tilde{\theta}_0, \tilde{\tau}_0$ respectively.

We must verify conditions (a) and (b) of Definition 1.2.2. Since (d', h') is a higher matrix factorization, we need only check

$$dh_c \equiv f_c \, \mathrm{Id}_{A_0(c)} \bmod (f_1, \dots, f_{c-1}) A_0(c)$$
$$\pi_c h_c d \equiv f_c \pi_c \bmod (f_1, \dots, f_{c-1}) B_1(c) \,.$$

Condition (a) holds because

$$\begin{pmatrix} d' & \psi \\ 0 & b_c \end{pmatrix} \begin{pmatrix} \theta_2 & \tau_0 \\ \partial_2 & \theta_0 \end{pmatrix} = \begin{pmatrix} d'\theta_2 + \theta_1 \partial_2 & d'\tau_0 + \theta_1 \theta_0 \\ \partial_1 \partial_2 & \partial_1 \theta_0 \end{pmatrix} \equiv \begin{pmatrix} f_c & 0 \\ 0 & f_c \end{pmatrix}$$

by (6.11). Similarly, Condition (b) is verified by the computation:

$$\begin{pmatrix} \theta_2 & \tau_0 \\ \partial_2 & \theta_0 \end{pmatrix} \begin{pmatrix} d' & \psi \\ 0 & b_c \end{pmatrix} = \begin{pmatrix} \theta_2 d' & \theta_2\theta_1 + \tau_0\partial_1 \\ \partial_2 d' & \partial_2\theta_1 + \theta_0\partial_1 \end{pmatrix} \equiv \begin{pmatrix} * & * \\ 0 & f_c \end{pmatrix}.$$

Next we show that the higher matrix factorization that we have constructed is pre-stable. Consider the complex (6.9), which is a free resolution of L over R'. It follows that

$$\mathrm{Coker}\Big(\widetilde{A}_0(c-1) \xrightarrow{\tilde{\partial}_2} \widetilde{B}_1(c)\Big) \cong \mathrm{Im}(\tilde{\partial}_1) \subset \widetilde{B}_0(c)$$

has no f_c-torsion, verifying the pre-stability condition.

It remains to show that d and h are liftings to S of the first two differentials in the minimal R-free resolution of M.

By (6.10) and Theorem 6.3.2 we have the following homotopies on the minimal R'-free resolution of M:

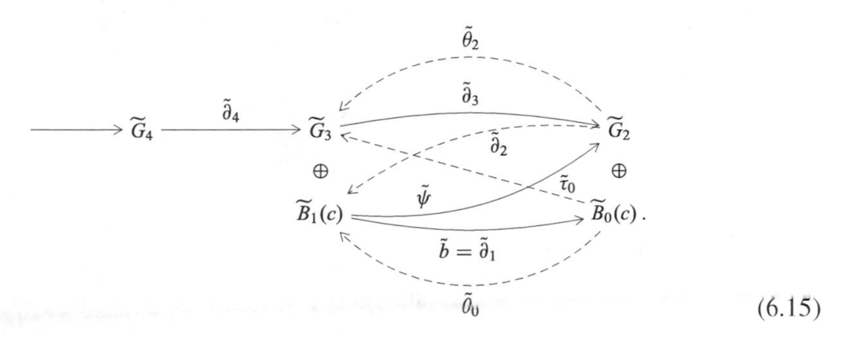

(6.15)

The minimal R-free resolution of M is obtained from the resolution above by applying the Shamash construction. Hence, the first two differentials are:

$$R \otimes \begin{pmatrix} \tilde{\partial}_3 & \tilde{\psi} \\ 0 & \tilde{b} \end{pmatrix} \quad \text{and} \quad R \otimes \begin{pmatrix} \tilde{\partial}_4 & \widetilde{\theta_2} & \tau_0 \\ 0 & \tilde{\partial}_2 & \widetilde{\theta_0} \end{pmatrix}.$$

By induction hypothesis $\tilde{\partial}_3 = R' \otimes d_{c-1}$ and $\tilde{\partial}_4 = R' \otimes h(c-1)$. By the construction of d and h in (6.13), (6.14) we see that $R \otimes d$ and $R \otimes h$ are the first two differentials in the minimal R-free resolution of M.

Finally, we will prove that if M is a stable syzygy, then (d, h) is stable as well. The map ∂_2 is the composite map

$$A_0(p-1) \hookrightarrow A_0(p) \xrightarrow{h_p} A_1(p) \xrightarrow{\pi_p} B_1(p)$$

by construction (6.14). By (6.9) it follows that if L is a maximal Cohen-Macaulay R-module, then $\mathrm{Coker}(\tilde{\partial}_2)$ is a maximal Cohen-Macaulay R'-module, verifying the stability condition for a higher matrix factorization over $R(p-1)$. By induction, it follows that (d, h) is stable. □

Remark 6.4.4 In order to capture structure of high syzygies without minimality, Definition 6.1.1 can be modified as follows. We extend the definition of syzygies to non-minimal free resolutions: if $(\mathbf{F}, \delta)$ is an R-free resolution of an R-module P, then we define $\mathrm{Syz}_{i,\mathbf{F}}(P) = \mathrm{Im}(\delta_i)$. Suppose that $f_1, \ldots, f_c$ is a regular sequence in a local ring S, and set $R = S/(f_1, \ldots, f_c)$. Let $(\mathbf{F}, \delta)$ be an R-free resolution, and let $M = \mathrm{Im}(\delta_r)$ for a fixed $r \geq 2c$.

We say that M is a *pre-stable syzygy in* $\mathbf{F}$ with respect to $f_1, \ldots, f_c$ if either $c = 0$ and $M = 0$, or $c \geq 1$ and there exists a lifting $(\widetilde{\mathbf{F}}, \tilde{\delta})$ of (F, δ) to $R' = S/(f_1, \ldots, f_{c-1})$ such that the CI operator $\tilde{t}_c := (1/f_c)\tilde{\delta}^2$ is surjective and, setting $(\widetilde{\mathbf{G}}, \tilde{\partial}) := \mathrm{Ker}(\tilde{t}_c)$, the module $\mathrm{Im}(\tilde{\partial}_r)$ is pre-stable in $\widetilde{\mathbf{G}}_{\geq 2}$ with respect to $f_1, \ldots, f_{c-1}$.

With minor modifications, the proof of Theorem 6.4.2 yields the following result: Let $\mathbf{F}$ be an R-free resolution. If M is a pre-stable r-th syzygy in $\mathbf{F}$ with respect to $f_1, \ldots, f_c$ then M is the HMF module of a pre-stable matrix factorization (d, h) such that d and h are liftings to S of the consecutive differentials δ_{r+1} and δ_{r+2} in $\mathbf{F}$. If $\mathbf{F}$ is minimal then the higher matrix factorization is minimal.

We can use the concept of pre-stable syzygy and Proposition 6.3.3 in order to build the minimal free resolutions of the modules $\mathrm{Coker}(R(p-1) \otimes b_p)$: For each q, let $\mathbf{T}(q)$ be the minimal $R(q)$-free resolution

$$\cdots \longrightarrow T(q)_3 \longrightarrow R(q) \otimes \Big(\oplus_{i \leq q} A_0(i) \Big) \longrightarrow R(q) \otimes A_1(q) \xrightarrow{d_q} R(q) \otimes A_0(q)$$

of $M(q)$ from Construction 5.1.1 and Theorem 5.1.2.

Proposition 6.4.5 *Let (d, h) be a minimal pre-stable matrix factorization for a regular sequence $f_1, \ldots, f_c$ in a local ring S, and use the notation of* 1.4.1. *For every $1 \leq p \leq c$, set*

$$N(p) = \mathrm{Coker}(R(p-1) \otimes b_p).$$

Then

$$N(p) = \mathrm{Coker}(R(p) \otimes b_p).$$

With the resolution $\mathbf{T}(q)$ defined as above, the minimal $R(p-1)$-free resolution $\mathbf{V}(p-1)$ of $N(p)$ is obtained by adjoining two maps to the right of $\mathbf{T}(p-1)$:

$$\mathbf{V}(p-1): \ \cdots \longrightarrow T(p-1)_0 \xrightarrow{\partial} R(p-1) \otimes B_1(p) \xrightarrow{R(p-1) \otimes b_p} R(p-1) \otimes B_0(p),$$

where ∂ is induced by the composite map

$$A_0(p-1) \hookrightarrow A_0(p) \xrightarrow{h_p} A_1(p) \xrightarrow{\pi_p} B_1(p)\,.$$

The minimal $R(p)$-free resolution of $N(p)$ is:

$$\mathbf{W}(p): \quad \cdots \longrightarrow T(p)_0 \xrightarrow{\delta} R(p) \otimes B_1(p) \xrightarrow{R(p)\otimes b_p} R(p) \otimes B_0(p)\,,$$

where

$$T(p)_0 = R(p) \otimes A_0(p) = \Big(R(p) \otimes B_0(p)\Big) \oplus \Big(R(p) \otimes T(p-1)_0\Big)$$

and δ is given by the Shamash construction applied to $\mathbf{V}(p-1)_{\leq 3}$.

Proof By Theorem 5.1.2 (using the notation in that theorem) the complex $\mathbf{T}(p)$ is an $R(p)$-free resolution of $M(p)$. By Theorem 5.4.4, the minimal $R(p-1)$-free resolution of $M(p)$ is:

$$\begin{array}{ccccc} \longrightarrow T(p-1)_2 \longrightarrow & T(p-1)_1 & \xrightarrow{\quad R(p-1)\otimes d_{p-1} \quad} & & T(p-1)_0 \\ & \oplus & \nearrow_{R(p-1)\otimes \psi_p} & & \oplus \\ & R(p-1)\otimes B_1(p) & \xrightarrow{\quad R(p-1)\otimes b_p \quad} & & R(p-1)\otimes B_0(p)\,. \end{array}$$

Since f_p is a non-zerodivisor on $M(p-1)$ by Corollary 3.2.3 and since the matrix factorization is pre-stable, we can apply Proposition 6.3.3, where the homotopies θ_i and τ_i for f_p are chosen to be the appropriate components of the map $R(p-1) \otimes h_p$. We get the minimal $R(p-1)$-free resolution

$$\mathbf{V}(p-1): \quad \mathbf{T}(p-1) \longrightarrow R(p-1) \otimes B_1(p) \xrightarrow{R(p-1)\otimes b_p} R(p-1) \otimes B_0(p)\,,$$

where the second differential is induced by the composite map

$$A_0(p-1) \hookrightarrow A_0(p) \xrightarrow{h_p} A_1(p) \xrightarrow{\pi_p} B_1(p)\,.$$

Since we have a homotopy for f_p on

$$R(p-1) \otimes B_1(p) \longrightarrow R(p-1) \otimes B_0(p)$$

it follows that $N(p) = \text{Coker}(R(p) \otimes b_p)\,.$

We next apply the Shamash construction to the following diagram with homotopies:

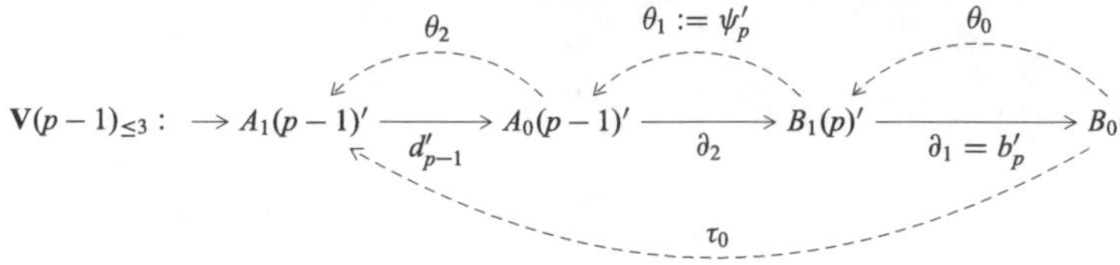

$$\mathbf{V}(p-1)_{\le 3}: \longrightarrow A_1(p-1)' \xrightarrow{d'_{p-1}} A_0(p-1)' \xrightarrow{\partial_2} B_1(p)' \xrightarrow{\partial_1 = b'_p} B_0(p)',$$

with homotopies θ_2, $\theta_1 := \psi'_p$, θ_0 and τ_0,

where $-'$ stands for $R(p-1) \otimes -$. By Proposition 4.3.2 we obtain an exact sequence

$$R(p) \otimes A_1(p) \longrightarrow R(p) \otimes A_0(p) \longrightarrow R(p) \otimes B_1(p) \longrightarrow R(p) \otimes B_0(p).$$

It is minimal since θ_0 is induced by h_p. The leftmost differential

$$R(p) \otimes A_1(p) \xrightarrow{R(p)\otimes b_p} R(p) \otimes A_0(p)$$

coincides with the first differential in $\mathbf{T}(p)$. □

6.5 Betti Numbers of Pre-stable Matrix Factorizations

Using Propositions 4.2.6 and 6.4.5, we can strengthen Corollaries 3.2.7 and 5.2.3 for pre-stable matrix factorizations:

Corollary 6.5.1 *Let S be local and (d, h) be a minimal pre-stable matrix factorization, and use the notation of* 1.4.1. *Let γ be the minimal number such that $A(\gamma) \neq 0$, so $\mathrm{cx}_R(M) = c - \gamma + 1$ by Corollary* 5.2.3. *Then*

$$\mathrm{rank}\,(B_1(p)) > \mathrm{rank}\,(B_0(p)) > 0 \quad \textit{for every } \gamma + 1 \le p \le c.$$

Proof Let $\gamma + 1 \le p \le c$. By Corollary 5.2.3, $\mathrm{rank}\,(B_1(p)) \neq 0$. Apply Proposition 6.4.5. Since the free resolution $\mathbf{V}(p-1)$ is minimal, it follows that $B_0(p) \neq 0$. The inequality $\mathrm{rank}\,(B_1(p)) > \mathrm{rank}\,(B_0(p))$ is established in Corollary 3.2.7. □

Remark 6.5.2 Under the assumptions and notation in 6.5.1, combining Corollaries 5.2.3 and 6.5.1, we get:

$$\begin{aligned}
&\mathrm{rank}\,(B_1(p)) = \mathrm{rank}\,(B_0(p)) = 0 \quad \text{for every } 1 \le p \le \gamma - 1\\
&\mathrm{rank}\,(B_1(\gamma)) = \mathrm{rank}\,(B_0(\gamma)) > 0\\
&\mathrm{rank}\,(B_1(p)) > \mathrm{rank}\,(B_0(p)) > 0 \quad \text{for every } \gamma + 1 \le p \le c.
\end{aligned}$$

It follows at once that the higher matrix factorization in Example 3.2.8 is not pre-stable.

Corollary 6.5.1 implies stronger restrictions on the Betti numbers in the finite resolution of modules that are pre-stable syzygies, for example:

Corollary 6.5.3 *If M is a pre-stable syzygy of complexity ζ with respect to the regular sequence $f_1, \ldots, f_c$ in a local ring S and $\beta_i^S(M)$ denotes the i-th Betti number of M as an S-module, then:*

$$\beta_0^S(M) \geq \zeta$$

$$\beta_1^S(M) \geq (c - \zeta + 1)\beta_0^S(M) + \binom{\zeta}{2}.$$

Proof Set $\gamma = c - \zeta + 1$. By Theorems 3.1.4, 5.1.2, and Corollary 6.5.1 we get

$$\beta_0^S(M) = \beta_0^{R(\gamma)}(M) = \sum_{p=\gamma}^{c} \text{rank } B_0(p) \geq \zeta$$

and

$$\beta_1^S(M) = \beta_1^{R(\gamma)}(M) + (c - \zeta)\beta_0^{R(\gamma)}(M) = \beta_1^{R(\gamma)}(M) + (c - \zeta)\beta_0^S(M)$$

$$\begin{aligned}
\beta_1^{R(\gamma)}(M) &= \sum_{p=\gamma}^{c} \text{rank } B_1(p) + \sum_{p=\gamma+1}^{c} (p - 1 - \gamma)\text{rank } B_0(p) \\
&\geq \left(c - \gamma + 1 - 1 + \sum_{p=\gamma}^{c} \text{rank } B_0(p)\right) + \sum_{p=\gamma+1}^{c} (p - 1 - \gamma)\text{rank } B_0(p) \\
&= \zeta - 1 + \beta_0^S(M) + \sum_{p=\gamma+1}^{c} (p - 1 - \gamma)\text{rank } B_0(p)\,.
\end{aligned}$$

Therefore,

$$\begin{aligned}
\beta_1^S(M) &\geq (c - \zeta + 1)\beta_0^S(M) + \zeta - 1 + \sum_{p=\gamma+1}^{c} (p - 1 - \gamma)\text{rank } B_0(p) \\
&\geq (c - \zeta + 1)\beta_0^S(M) + \zeta - 1 + \binom{\zeta - 1}{2} \\
&= (c - \zeta + 1)\beta_0^S(M) + \binom{\zeta}{2}.
\end{aligned}$$

□

For example, a pre-stable syzygy module of complexity ≥ 2 cannot be cyclic and cannot have $\beta_1^S(M) = \beta_0^S(M) + 1$.

Chapter 7
The Gorenstein Case

Abstract In this chapter we consider the case when S is a local Gorenstein ring.

7.1 Syzygies and Maximal Cohen-Macaulay Modules

In this section we review some well-known results on maximal Cohen-Macaulay syzygies.

Lemma 7.1.1 *Let R be a local Cohen-Macaulay ring.*

(1) *If N is an R-module, then for $j \geq \operatorname{depth}(R) - \operatorname{depth}(N)$, the R-module $\operatorname{Syz}_j^R(N)$ is a maximal Cohen-Macaulay module.*
(2) *If M is the first syzygy module in a minimal free resolution of a maximal Cohen-Macaulay R-module, then M has no free summands.*

Proof (1) follows immediately from Lemma 6.1.12. We prove (2). Let $M = \operatorname{Syz}_1(N)$. Factoring out a maximal N-regular and R-regular sequence, we reduce to the artinian case. Suppose R is artinian. Let $\mathbf{F}$ be the minimal free resolution of N. Since $\mathbf{F}$ is minimal, $\operatorname{Syz}_1^R(N)$ is contained in the maximal ideal times $F_0 \otimes R$. Hence, it is annihilated by the socle of R, and cannot contain a free submodule. □

We now assume that R has a canonical module ω_R—this is the case whenever R is a homomorphic image of a Gorenstein local ring S, in which case $\omega_R = \operatorname{Ext}^{\dim S - \dim R}(R, S)$. In this case a nonzero R-module M is a maximal Cohen-Macaulay module if and only if $\operatorname{Ext}_R^i(M, \omega_R) = 0$ for all $i > 0$. For the rest of this section we will work over a local Gorenstein ring.

Lemma 7.1.2 *Let R be a local Gorenstein ring. If M is a maximal Cohen-Macaulay R-module, then M^* is a maximal Cohen-Macaulay R-module, M is reflexive, and $\operatorname{Ext}_R^i(M, R) = 0$ for all $i > 0$.*

A proof of Lemma 7.1.2 can be found for example in [26, Proposition 21.12] and [26, Theorem 21.21].

Over a local Gorenstein ring matters are simplified by the fact that a maximal Cohen-Macaulay module is, in a canonical way, an m-th syzygy for any m by the next lemma. When M is a maximal Cohen-Macaulay R-module without free summands we let $\operatorname{Syz}_{-j}^R(M)$ be the dual of the j-th syzygy of $M^* := \operatorname{Hom}_R(M, R)$.

D. Eisenbud, I. Peeva, *Minimal Free Resolutions over Complete Intersections*,
Lecture Notes in Mathematics 2152, DOI 10.1007/978-3-319-26437-0_7

When we speak of syzygies or cosyzygies, we will implicitly suppose that they are taken with respect to a minimal resolution.

Lemma 7.1.3 *Let R be a local Gorenstein ring.*

(1) *The functor* $\operatorname{Hom}_R(-, R)$ *takes exact sequences of maximal Cohen-Macaulay modules to exact sequences.*
(2) *If M is a maximal Cohen-Macaulay module without free summands, then there exists a unique maximal Cohen-Macaulay R-module $N := \operatorname{Syz}^R_{-j}(M)$ without free summands such that M is isomorphic to $\operatorname{Syz}^R_j(N)$.*
(3) *If M is a maximal Cohen-Macaulay module without free summands, then*

$$M \cong \operatorname{Syz}^R_j(\operatorname{Syz}^R_{-j}(M)) \cong \operatorname{Syz}^R_{-j}(\operatorname{Syz}^R_j(M))$$

for every $j \geq 0$.

Proof (2) and (3) follow from (1) and Lemma 7.1.2. We prove (1). A short exact sequence $0 \longrightarrow M \longrightarrow M' \longrightarrow M'' \longrightarrow 0$ yields a long exact sequence

$$0 \longrightarrow \operatorname{Hom}_R(M'', R) \longrightarrow \operatorname{Hom}_R(M', R) \longrightarrow \operatorname{Hom}_R(M, R) \longrightarrow \operatorname{Hom}_R(-, R)$$
$$\longrightarrow \operatorname{Ext}^1_R(M'', R) \longrightarrow \cdots,$$

and $\operatorname{Ext}^1_R(M'', R) = 0$ by Lemma 7.1.2. □

7.2 Stable Syzygies in the Gorenstein Case

In this section S will denote a local Gorenstein ring. We write $f_1, \dots, f_c$ for a regular sequence in S, and thus $R = S/(f_1, \dots, f_c)$ is also a Gorenstein ring. We will show that stable syzygies all come from stable matrix factorizations.

Theorem 7.2.1 *Let $f_1, \dots, f_c$ be a regular sequence in a Gorenstein local ring S, and set $R = S/(f_1, \dots, f_c)$. An R-module M is a stable syzygy if and only if it is the module of a minimal stable matrix factorization with respect to $f_1, \dots, f_c$.*

Recall that in Proposition 6.4.5 we showed that

$$\operatorname{Coker}(R(p) \otimes b_p) = \operatorname{Coker}(R(p-1) \otimes b_p)$$

and defined this to be $N(p)$; we also gave free resolutions $\mathbf{V}(p-1)$ and $\mathbf{W}(p)$ of $N(p)$ as an $R(p-1)$-module and as an $R(p)$-module, respectively. We postpone the proof of Theorem 7.2.1 to emphasize the relationship between $M(p), N(p)$ and $M(p-1)$.

Proposition 7.2.2 *Let $f_1, \dots, f_c$ be a regular sequence in a Gorenstein local ring S, and set $R = S/(f_1, \dots, f_c)$. Let M be the HMF module of a minimal stable matrix* *factorization* (d,h) *for* $f_1,\dots,f_c$. *With notation as above,*

$$N(p) = \mathrm{Syz}_{-2}^{R(p)}(M(p))$$

$$M(p-1) = \mathrm{Syz}_{2}^{R(p-1)}\Big(N(p)\Big),$$

and thus

$$M(p-1) \cong \mathrm{Syz}_{2}^{R(p-1)}\Big(\mathrm{Syz}_{-2}^{R(p)}\Big(M(p)\Big)\Big).$$

Proof We apply Proposition 6.4.5. As the higher matrix factorization is stable, we conclude that the depth of the $R(p-1)$-module $\mathrm{Coker}(R(p-1)\otimes b_p)$ is at most one less than that of a maximal Cohen-Macaulay $R(p-1)$-module. Therefore, $N(p)$ is a maximal Cohen-Macaulay $R(p)$-module. It has no free summands because the map b_p is part of a higher matrix factorization. Note that $\mathrm{Syz}_2^{R(p)}(N(p)) = M(p)$ by the free resolution $\mathbf{W}$ in Proposition 6.4.5.

Furthermore,

$$M(p-1) = \mathrm{Coker}(R(p-1)\otimes d_{p-1}) = \mathrm{Syz}_{2}^{R(p-1)}\Big(N(p)\Big),$$

where the last equality follows from the free resolution $\mathbf{V}$ in Proposition 6.4.5. □

Proof of Theorem 7.2.1 Theorem 6.4.2 shows that a stable syzygy yields a stable matrix factorization.

Conversely, let M be the module of a minimal stable matrix factorization (d,h). Use notation as in 1.4.1. By Propositions 6.4.5 and 7.2.2 and their notation, $\mathbf{W}(p)$ is the minimal R-free resolution of

$$\mathrm{Syz}_{-2}^{R(p)}(M(p)) = \mathrm{Coker}(R(p)\otimes b_p).$$

We have a surjective CI operator t_c on $\mathbf{W}(p)$ because on the one hand, we have it on $\mathbf{T}(p)$ and on the other hand $\mathbf{W}(p)_{\leq 3}$ is given by the Shamash construction so we have a surjective standard CI operator on $\mathbf{W}(p)_{\leq 3}$. Furthermore, the standard lifting of $\mathbf{W}(p)$ to $R(p-1)$ starts with $\mathbf{V}(p-1)_{\leq 1}$, so in the notation of Definition 6.1.1 we get $\mathrm{Ker}(\tilde{\delta}_1) = M(p-1)$, which is stable by induction. □

7.3 Maximal Cohen-Macaulay Approximations

We will next show that the sequence of the intermediate modules $M(p)$ of a stable matrix factorization is a sequence of Cohen-Macaulay approximations in the sense of Auslander and Buchweitz [3], and use this idea to give an alternate proof of

Theorem 6.1.7 in the case when S is Gorenstein, showing that high syzygies are stable syzygies.

Throughout the section S will denote a local Gorenstein local ring with residue field k. The main object in this section is defined as follows:

Definition 7.3.1 Let N be an S-module. Choose an integer $q > \operatorname{depth} S - \operatorname{depth} N$ and set

$$\operatorname{App}_S(N) := \operatorname{Syz}^S_{-q}\left(\operatorname{Syz}^S_q(N)\right),$$

which we will call the *essential CM S-approximation of N*. By Lemmas 7.1.1 and 7.1.3, $\operatorname{App}_S(N)$ is a maximal Cohen-Macaulay module without free summands, independent of the choice of $q > \operatorname{depth} S - \operatorname{depth} N$.

Remark 7.3.2 The theory of Cohen-Macaulay approximations, introduced by Auslander and Buchweitz [3], is quite general, but it takes a simple form for modules over a local Gorenstein ring (see also [20, Sect. 1.1, pp. 14–16]). The Cohen-Macaulay approximation of an S-module N is, by definition, a surjective map to N from a maximal Cohen-Macaulay module having a certain universal property. It is the direct sum of a free S-module and a module without free summands. It is easy to see that the latter is $\operatorname{App}_S(N)$, cf. [20].

In what follows we deal with modules that are naturally modules over several different rings at once. Clearly, for any $i \geq 1$:

$$\operatorname{App}_S\left(\operatorname{Syz}^S_i(N)\right) = \operatorname{Syz}^S_i\left(\operatorname{App}_S(N)\right).$$

We extend this to taking syzygies over other rings:

Theorem 7.3.3 *Let $R = S/I$ be a factor ring of the local Gorenstein ring S, and suppose that R has finite projective dimension as an S-module. Let N be an R-module.*

(1) *For any $i \geq 0$,*

$$\operatorname{App}_S\left(\operatorname{Syz}^R_i(N)\right) = \operatorname{Syz}^S_i\left(\operatorname{App}_S(N)\right).$$

If N is a maximal Cohen-Macaulay module without free summands, then the statement is also true for $i < 0$.

(2) *If $j > \operatorname{depth} S - \operatorname{depth} N$, then*

$$\operatorname{App}_S\left(\operatorname{Syz}^R_j(N)\right) = \operatorname{Syz}^S_j(N).$$

Proof

(1) It suffices to do the cases of first syzygies and cosyzygies. Let

$$0 \longrightarrow N' \longrightarrow F \longrightarrow N \longrightarrow 0$$

be a short exact sequence, with F free as an R-module. It suffices to show that $\mathrm{Syz}_i^S(N') = \mathrm{Syz}_{i+1}^S(N)$ for some i.

We may obtain an S-free resolution of N as the mapping cone of the induced map from the S-free resolution of N' to the S-free resolution of F; if the projective dimension of R as an S-module (and thus of F as an S-module) is u, it follows that $\mathrm{Syz}_{u+1}^S(N') = \mathrm{Syz}_u^S(N)$.

(2) Apply part(1) and note that

$$\mathrm{Syz}_j^S\Big(\mathrm{App}_S(N)\Big) = \mathrm{Syz}_j^S(N)$$

for $j > \operatorname{depth} S - \operatorname{depth} N$. □

Corollary 7.3.4 *Let $R = S/I$ be a factor ring of the local Gorenstein ring S, and suppose that R has finite projective dimension as an S-module. If two R-modules N and N' have a common R-syzygy $\mathrm{Syz}_i^R(N) = \mathrm{Syz}_i^R(N')$ then*

$$\mathrm{App}_S(N) = \mathrm{App}_S(N')\,.$$

In particular,

$$\mathrm{App}_S\Big(\mathrm{App}_R(N)\Big) = \mathrm{App}_S(N)\,.$$

Proof By Theorem 7.3.3,

$$\begin{aligned}\mathrm{Syz}_i^S(\mathrm{App}_S(N)) &= \mathrm{App}_S(\mathrm{Syz}_i^R(N)) \\ &= \mathrm{App}_S(\mathrm{Syz}_i^R(N')) \\ &= \mathrm{Syz}_i^S(\mathrm{App}_S(N'))\,.\end{aligned}$$

Since $\mathrm{App}_S(N)$ and $\mathrm{App}_S(N')$ have a common syzygy over S, we conclude $\mathrm{App}_S(N) = \mathrm{App}_S(N')$.

The last statement follows because $\mathrm{App}_R(N)$ and N have a common R-syzygy. □

We can use Theorem 7.4.1 to give another proof that a sufficiently high syzygy of any module N over a complete intersection $R = S/(f_1, \ldots, f_c)$ is stable. The main idea is that each $\mathrm{App}_{R(p)}\Big(\mathrm{Syz}_v^R(N)\Big)$ is a v-th syzygy over $R(p)$, and thus all of these become high syzygies over their own rings when v is large.

Corollary 7.3.5 *Let $f_1, \dots, f_c$ be a regular sequence in S, set $R(p) = S/(f_1, \dots, f_c)$, and let N be an $R = R(c)$-module. Denote $\mathcal{R}(p) = k[\chi_1, \dots, \chi_p]$ the ring of CI operators corresponding to $f_1, \dots, f_p$. There exists an integer υ_0 depending on N such that, for any $\upsilon \geq \upsilon_0$, the $R(p)$-module $\mathrm{Syz}_\upsilon^{R(p)}(N)$ is maximal Cohen-Macaulay without free summands and*

$$\mathrm{reg}_{\mathcal{R}(p)}\Big(\mathrm{Ext}_{R(p)}\Big(\mathrm{App}_{R(p)}(\mathrm{Syz}_\upsilon^R(N)),\ k\Big)\Big) = 1$$

for all $1 \leq p \leq c$. Thus,

$$\mathrm{depth}_{\mathcal{R}(p)}\Big(\mathrm{Ext}_{R(p)}\Big(\mathrm{App}_{R(p)}(\mathrm{Syz}_{\upsilon+2}^R(N)),\ k\Big)\Big) > 0\,.$$

Proof By Theorem 7.3.3 we have

$$\mathrm{App}_{R(p)}\Big(\mathrm{Syz}_\upsilon^R(N)\Big) = \mathrm{Syz}_\upsilon^{R(p)}\Big(\mathrm{App}_{R(p)}(N)\Big)\,,$$

so $\mathrm{Ext}_{R(p)}\Big(\mathrm{App}_{R(p)}(\mathrm{Syz}_\upsilon^R(N)),\ k\Big)$ is a truncation of $\mathrm{Ext}_{R(p)}\Big(\mathrm{App}_{R(p)}(N),\ k\Big)$. Therefore, we can make its regularity smaller by choosing υ larger. By our conventions, 1 is the minimal possible value of the regularity of an ext-module.

Set

$$E(p) = \mathrm{Ext}_{R(p)}\Big(\mathrm{App}_{R(p)}(\mathrm{Syz}_\upsilon^R(N)),\ k\Big)\,.$$

The maximal submodule of $E(p)$ of finite length is contained in $E(p)^{\leq 1}$ since $\mathrm{reg}(E(p)) = 1$. We conclude that the maximal ideal $(\chi_1, ..., \chi_p)$ is not an associated prime of $E(p)^{\geq 2}$, so its depth is positive. □

Corollary 7.3.6 *Suppose that the residue field of the local ring S is infinite. With notation as in Corollary 7.3.5, suppose that $j \geq \upsilon_0 + 4$. If $f'_1, \dots, f'_c$ is a sequence of generators of $(f_1, \dots, f_c)$ that is generic for N, then $M := \mathrm{Syz}_j^R(N)$ is a stable syzygy with respect to $f'_1, \dots, f'_c$.*

Proof For each $p = 1, \dots, c$ we set

$$M(p) = \mathrm{App}_{R(p)}(M)$$
$$L(p) = \mathrm{Syz}_{-2}^{R(p)}(M(p))\,.$$

Since $L(p)$ is a maximal Cohen-Macaulay $R(p)$-module, by Part (2) of Theorem 7.3.3,

$$\mathrm{Syz}_2^{R(p-1)}\Big(L(p)\Big) = \mathrm{App}_{R(p-1)}\Big(\mathrm{Syz}_2^{R(p)}(L(p))\Big) = \mathrm{App}_{R(p-1)}\Big(M(p)\Big)\,.$$

Furthermore,

$$\mathrm{App}_{R(p-1)}(M(p)) = \mathrm{App}_{R(p-1)}\Big(\mathrm{App}_{R(p)}(M)\Big)$$
$$= \mathrm{App}_{R(p-1)}(M) = M(p-1)\,,$$

where the second equality holds by Corollary 7.3.4.

We showed that $M(p-1) = \mathrm{Syz}_2^{R(p-1)}\Big(L(p)\Big)$, and by the definition of $L(p)$ we have $M(p) = \mathrm{Syz}_2^{R(p)}\Big(L(p)\Big)$. Apply Corollaries 7.3.5 and 4.3.5 to conclude that the CI operator corresponding to f_p' is surjective on the minimal $R(p)$-free resolution of $L(p)$. Thus M is a stable syzygy. □

7.4 Stable Matrix Factorizations over a Gorenstein Ring

In this section we suppose that S is a local Gorenstein ring, and derive some corollaries of the results in the previous section.

The following result identifies in two different ways the modules $M(p)$ as coming from a single module over R.

Theorem 7.4.1 *Let S be a Gorenstein local ring, and suppose that $M = M(c)$ is the module over $R(c) = S/(f_1, \dots, f_c)$ of a stable minimal matrix factorization of a regular sequence $f_1, \dots, f_c$. We use notation as in Sect. 1.4.*

(1) *For every $p = 1, \dots, c$ we have:*

$$M(p) = \mathrm{App}_{R(p)}(M)\,.$$

(2) *Suppose that $M = \mathrm{Syz}_j^R(N)$ for some R-module N with $j > \mathrm{depth}\, S - \mathrm{depth}\, N$. For every $p = 1, \dots, c$ we have:*

$$M(p) = \mathrm{Syz}_j^{R(p)}(N) \ \ \textit{for}\ p > 0\,.$$

In particular, if we set $P = \mathrm{Syz}_{-c-1}^R(M)$ then

$$M(p) = \mathrm{Syz}_{c+1}^{R(p)}(P)\,.$$

Proof

(1) The proof is by induction on p. Let $N(p)$ be the modules defined in Proposition 6.4.5. By Corollary 7.2.2,

$$M(p) = \mathrm{Syz}_2^{R(p)}\Big(N(p)\Big)$$
$$M(p-1) = \mathrm{Syz}_2^{R(p-1)}\Big(N(p)\Big)\,.$$

Part (2) of Theorem 7.3.3 shows that $M(p-1) = \mathrm{App}_{R(p-1)}\Big(M(p)\Big)$. Using Corollary 7.3.4 and a descending induction on p, it follows that $M(p-1) = \mathrm{App}_{R(p-1)}(M)$ as required.

(2) The R-module M is maximal Cohen-Macaulay without free summands by Corollaries 3.2.2 and 3.2.5. Hence,

$$M(p) = \mathrm{App}_{R(p)}(M) = \mathrm{App}_{R(p)}\Big(\mathrm{Syz}_j^R(N)\Big) = \mathrm{Syz}_j^{R(p)}(N)$$

by Theorem 7.3.3(2). □

Since syzygies of stable HMF modules are again stable HMF modules by Proposition 6.1.5, it makes sense to ask about the intermediate modules associated to them:

Proposition 7.4.2 *Under the assumptions and in the notation of Theorem* 7.4.1, *for every* $p = 1, \dots, c$ *we have:*

$$\Big(\mathrm{Syz}_1^R(M)\Big)(p) = \mathrm{Syz}_1^{R(p)}\Big(M(p)\Big).$$

Proof Applying Theorem 7.4.1 and part (1) of Theorem 7.3.3 we get

$$\begin{aligned}\Big(\mathrm{Syz}_1^R(M)\Big)(p) &= \mathrm{App}_{R(p)}(\mathrm{Syz}_1^R(M)) \\ &= \mathrm{Syz}_1^{R(p)}\Big(\mathrm{App}_{R(p)}(M)\Big) = \mathrm{Syz}_1^{R(p)}(M(p)).\end{aligned}$$

□

Corollary 7.4.3 *Under the assumptions and notation of Theorem* 7.4.1, *let M be the module of a stable matrix factorization (d, h). If we denote the codimension* 1 *part of (d, h) by (d_1, h_1), then the codimension* 1 *part of the higher matrix factorization of $\mathrm{Syz}_1^R(M)$ is (h_1, d_1).*

Proof If (d_1, h_1) is non-trivial, then the minimal $R(1)$-free resolution of $M(1) = R(1) \otimes d_1$ is periodic of the form:

$$\cdots \xrightarrow{d_1} F_4 \xrightarrow{h_1} F_3 \xrightarrow{d_1} F_2 \xrightarrow{h_1} F_1 \xrightarrow{d_1} F_0.$$

□

Corollary 7.4.4 *Let $f_1, \dots, f_c$ be a regular sequence in a Gorenstein local ring S, and set $R = S/(f_1, \dots, f_c)$. Suppose that N is an R-module of finite projective dimension over S. Assume that $f_1, \dots, f_c$ are generic with respect to N. Choose a $j >$ depth S – depth N large enough so that $M := \mathrm{Syz}_j^R(N)$ is a stable syzygy. By Theorem* 7.2.1, *M is the module of a minimal stable matrix factorization with respect* *to* $f_1, \ldots, f_c$. *By Theorem* 7.4.1(2), *for every* $p = 1, \ldots, c$ *we have:*

$$M(p) = \mathrm{Syz}_j^{R(p)}(N) \ \ \textit{for}\, p \geq 0\,.$$

Denote

$$\gamma := c - \mathrm{cx}_R(N) + 1\,,$$

where $\mathrm{cx}_R(N)$ *is the complexity of* N *(see Corollary* 5.2.3*).*

(1) *The projective dimension of* N *over* $R(p) = S/(f_1, \ldots, f_p)$ *is finite for* $p < \gamma$.
(2) *The hypersurface matrix factorization for the periodic part of the minimal free resolution of* N *over* $S/(f_1, \ldots, f_\gamma)$ *is isomorphic to the top non-zero part of the higher matrix factorization of* M.

A version of (1) is proved in [4, Theorem 3.9], [7, 5.8 and 5.9].

Proof By 6.5.2, $M(p) = 0$ for $p < \gamma$. Apply Theorem 7.4.1(2). For $p < \gamma$ we establish (1). The case $p = \gamma$ establishes (2). □

Remark 7.4.5 In particular, Corollary 7.4.4 shows that the codimension 1 matrix factorization that is obtained from a high $S/(f_1)$-syzygy of N agrees with the codimension 1 part of the higher matrix factorization for M, and both codimension 1 matrix factorizations are trivial if the complexity of N is $< c$ (where $M = \mathrm{Syz}_j^R(N)$ is a stable syzygy of N with $j >$ depth $S -$ depth N).

Chapter 8
Functoriality

Abstract In this chapter, we introduce HMF morphisms, and show that any homomorphism of matrix factorization modules induces an HMF morphism.

8.1 HMF Morphisms

We introduce the concept of an HMF morphism (a morphism of higher matrix factorizations) so that it preserves the structures described in Definition 1.2.2:

Definition 8.1.1 A *morphism of matrix factorizations* or *HMF morphism* α : $(d, h) \longrightarrow (d', h')$ is a triple of homomorphisms of free modules

$$\alpha_0 : A_0 \longrightarrow A'_0$$

$$\alpha_1 : A_1 \longrightarrow A'_1$$

$$\alpha_2 : \oplus_{p \le c} A_0(p) \longrightarrow \oplus_{p \le c} A'_0(p)$$

such that, for each p:

(a) $\alpha_s(A_s(p)) \subseteq A'_s(p)$ for $s = 0, 1$. We write $\alpha_s(p)$ for the restriction of α_s to $A_s(p)$.
(b) $\alpha_2 \left(\oplus_{q \le p} A_0(q)\right) \subseteq \oplus_{q \le p} A'_0(q)$, and the component $A_0(p) \longrightarrow A'_0(p)$ of α_2 is $\alpha_0(p)$. We write $\alpha_2(p)$ for the restriction of α_2 to $\oplus_{q \le p} A_0(q)$.
(c) The diagram

$$\begin{array}{ccccc}
\oplus_{q\le p} A_0(q) & \xrightarrow{h} & A_1(p) & \xrightarrow{d_p} & A_0(p) \\
\Big\downarrow{\scriptstyle \alpha_2(p)} & & \Big\downarrow{\scriptstyle \alpha_1(p)} & & \Big\downarrow{\scriptstyle \alpha_0(p)} \\
\oplus_{q\le p} A'_0(q) & \xrightarrow{h'} & A'_1(p) & \xrightarrow{d'_p} & A'_0(p)
\end{array}$$

commutes modulo $(f_1, \dots, f_{p-1})$.

Theorem 8.1.2 *Suppose that $f_1, \dots, f_c$ is a regular sequence in a Gorenstein local ring S, and set $R = S/(f_1, \dots, f_c)$. Let M and M' be stable syzygies over R, and*

D. Eisenbud, I. Peeva, *Minimal Free Resolutions over Complete Intersections*, Lecture Notes in Mathematics 2152, DOI 10.1007/978-3-319-26437-0_8

suppose $\zeta : M \longrightarrow M'$ *is a morphisms of* R*-modules. With notation as in* 1.4.1, *let* M *and* M' *be HMF modules of stable matrix factorizations* (d,h) *and* (d',h'), *respectively. There exists an HMF morphism*

$$\alpha : (d,h) \longrightarrow (d',h')$$

such that the map induced on

$$M = \operatorname{Coker}(R \otimes d) \longrightarrow \operatorname{Coker}(R \otimes d') = M'$$

is ζ.

We first establish a strong functoriality statement for the Shamash construction. Suppose that $\mathbf{G}$ and $\mathbf{G}'$ are S-free resolutions of S-modules M and M' annihilated by a non-zerodivisor f, and $\zeta : M \longrightarrow M'$ is any homomorphism. If we choose systems of higher homotopies σ and σ' for f on $\mathbf{G}$ and $\mathbf{G}'$ respectively, then the Shamash construction yields resolutions $\operatorname{Sh}(\mathbf{G},\sigma)$ and $\operatorname{Sh}(\mathbf{G}',\sigma')$ of M and M over $R = S/(f)$, and thus there is a morphism of complexes

$$\tilde{\phi} : \operatorname{Sh}(\mathbf{G},\sigma) \longrightarrow \operatorname{Sh}(\mathbf{G}',\sigma')$$

covering ζ. To prove the Theorem we need more: a morphism defined over S that commutes with the maps in the "standard liftings" $\widetilde{\operatorname{Sh}}(\mathbf{G},\sigma)$ and $\widetilde{\operatorname{Sh}}(\mathbf{G}',\sigma')$ (see Construction 4.3.1) and respects the natural filtrations of these modules. The following statement provides the required morphism.

Lemma 8.1.3 *Let* S *be a commutative ring, and let* $\varphi_0 : (\mathbf{G},d) \longrightarrow (\mathbf{G}',d')$ *be a map of* S*-free resolutions of modules annihilated by an element* f. *Given systems of higher homotopies* σ_j *and* σ_j' *on* $\mathbf{G}$ *and* $\mathbf{G}'$, *respectively, there exists a system of maps* φ_j *of degree* $2j$ *from the underlying free module of* $\mathbf{G}$ *to that of* $\mathbf{G}'$ *such that, for every index* m,

$$\sum_{i+j=m} (\sigma_i'\varphi_j - \varphi_j\sigma_i) = 0.$$

We say that $\{\varphi_j\}$ is a *system of homotopy comparison maps* if they satisfy the conditions in the lemma above.

Recall that a map of free complexes $\lambda : \mathbf{U} \longrightarrow \mathbf{W}[-a]$ is a homotopy for a map $\rho : \mathbf{U} \longrightarrow \mathbf{W}[-a+1]$ if

$$\delta\lambda - (-1)^a \lambda\partial = \rho,$$

where ∂ and δ are the differentials in $\mathbf{U}$ and $\mathbf{W}$ respectively. Since in Lemma 8.1.3 σ_0 and σ_0' are the differentials d and d', the equation above in Lemma 8.1.3 says that, for each m, the map φ_m is a homotopy for the sum

$$-\sum_{\substack{i+j=m\\ i>0, j>0}} (\sigma_i'\varphi_j - \varphi_j\sigma_i).$$

Proof The desired condition on φ_0 is equivalent to the given hypothesis that φ_0 is a map of complexes. We proceed by induction on $m > 0$ and on homological degree to prove the existence of φ_m. The desired condition can be written as:

$$d'\varphi_m = -\sum_{\substack{i+j=m\\ i\neq 0}} \sigma'_i\varphi_j + \sum_{i+j=m} \varphi_j\sigma_i\,.$$

Since $\mathbf{G}$ is a free resolution, it suffices to show that the right-hand side is annihilated by d'. Indeed,

$$-\sum_{\substack{i+j=m\\ i\neq 0}} (d'\sigma'_i)\varphi_j + \sum_{i+j=m} (d'\varphi_j)\sigma_i$$

$$= \sum_{\substack{i+j=m\\ i\neq 0}}\sum_{\substack{v+w=i\\ v\neq 0}} \sigma'_v\sigma'_w\varphi_j - f\varphi_{m-1} - \sum_{i+j=m}\sum_{\substack{q+u=j\\ q\neq 0}} \sigma'_q\varphi_u\sigma_i$$

$$+ \sum_{i+j=m}\sum_{u+q=j} \varphi_u\sigma_q\sigma_i$$

$$= \sum_{\substack{v+w+j=m\\ v\neq 0}} \sigma'_v\sigma'_w\varphi_j - f\varphi_{m-1} - \sum_{\substack{i+q+u=m\\ q\neq 0}} \sigma'_q\varphi_u\sigma_i + \sum_{i+u+q=m} \varphi_u\sigma_q\sigma_i\,,$$

where the first equality holds by (3) in Definition 3.4.1 and by the induction hypothesis. Reindexing the first summand by $v = q$, $w = i$ and $j = u$ we get:

$$\sum_{\substack{q+i+u=m\\ q\neq 0}} \sigma'_q\sigma'_i\varphi_u - f\varphi_{m-1} - \sum_{\substack{i+q+u=m\\ q\neq 0}} \sigma'_q\varphi_u\sigma_i + \sum_{i+u+q=m} \varphi_u\sigma_q\sigma_i$$

$$= -f\varphi_{m-1} + \sum_{q\neq 0}\sigma'_u\left(\sum_{i+u=m-q}\sigma'_i\varphi_u - \varphi_u\sigma_i\right) + \sum_u \varphi_u\left(\sum_{q+i=m-u}\sigma_q\sigma_i\right)$$

$$= -f\varphi_{m-1} + 0 + 0 + \varphi_{m-1}f$$

$$= 0\,,$$

where the last equality holds by (3) in Definition 3.4.1 and by induction hypothesis. □

The next result reinterprets the conditions of Lemma 8.1.3 as defining a map between liftings of Shamash resolutions.

Proposition 8.1.4 *Let S be a commutative ring, and let $\mathbf{G}$ and $\mathbf{G}'$ be S-free resolutions with systems of higher homotopies $\sigma = \{\sigma_j\}$ and $\sigma' = \{\sigma'_j\}$ for*

$f \in S$, respectively. Suppose that $\{\varphi_j\}$ is a system of homotopy comparison maps for σ and σ'. We use the standard lifting of the Shamash resolution defined in Construction 4.3.1, *and the notation established there. Denote by $\tilde{\varphi}$ the map with components*

$$\varphi_i : y^{(v)}G_j \longrightarrow y^{(v-i)}G'_{j+2i}$$

from the underlying graded free S-module of the standard lifting $\widetilde{\mathrm{Sh}}(\mathbf{G}, \sigma)$ of the Shamash resolution $\mathrm{Sh}(\mathbf{G}, \sigma)$, *to the underlying graded free S-module of the standard lifting* $\widetilde{\mathrm{Sh}}(\mathbf{G}', \sigma')$ *of the Shamash resolution* $\mathrm{Sh}(\mathbf{G}', \sigma')$. *The maps $\tilde{\varphi}$ satisfy $\widetilde{\delta'}\tilde{\varphi} = \tilde{\varphi}\tilde{\delta}$, where $\tilde{\delta}$ and $\widetilde{\delta'}$ are the standard liftings of the differentials defined in Construction* 4.3.1.

Proof Fix a and v. We must show that the diagram

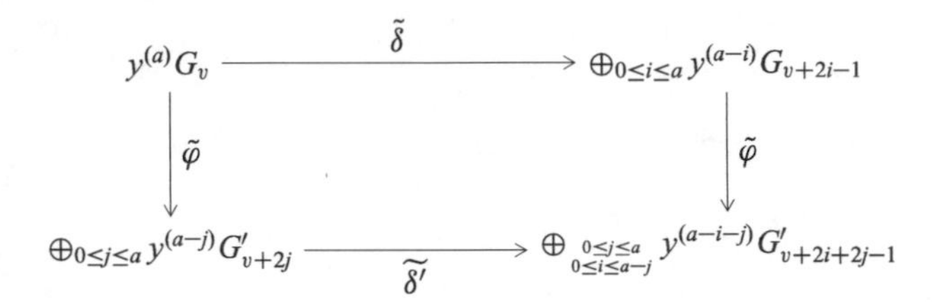

commutes. Fix $0 \le q \le a$. The map $\widetilde{\delta'}\tilde{\varphi} - \tilde{\varphi}\tilde{\delta}$ from $y^{(a)}G_v$ to $y^{(q)}G'_{v+2a-2q-1}$ is equal to

$$\sum_{i+j=a-q} (\sigma'_i\varphi_j - \varphi_j\sigma_i)\,,$$

which vanishes by Lemma 8.1.3. □

Remark 8.1.5 A simple modification of the proof of Lemma 8.1.3 shows that systems of homotopy comparison maps also exist in the context of systems of higher homotopies for a regular sequence $f_1, \ldots, f_c$, not just in the case $c = 1$ as above, and one can interpret this in terms of Shamash resolutions as in Proposition 8.1.4 as well, but we do not need these refinements.

Proof of Theorem 8.1.2 The result is immediate for $c = 1$, so we proceed by induction on $c > 1$. Let $\widetilde{R} = S/(f_1, \ldots, f_{c-1})$. To simplify the notation, we will write $\widetilde{\ }$ for $\widetilde{R} \otimes_S -$, and $\overline{\ }$ for $R \otimes -$. We will use the notation in 1.4.1.

Since (d, h) is stable we can extend the map $\overline{d}$ to a complex

$$\overline{A}_1(c) \longrightarrow \overline{A}_0(c) \longrightarrow \overline{B}_1(c) \longrightarrow \overline{B}_0(c)\,,$$

that is the beginning of an R-free resolution $\mathbf{F}$ of $\mathrm{Syz}^R_{-2}(M)$, and there is a similar complex that is the beginning of the R-free resolution $\mathbf{F}'$ of $\mathrm{Syz}^R_{-2}(M')$. By stability these cosyzygy modules are maximal Cohen-Macaulay modules, so dualizing these complexes we may use

$$\zeta(c) := \zeta : M \longrightarrow M'$$

to induce maps

$$\eta : \mathrm{Syz}^R_{-2}(M) \longrightarrow \mathrm{Syz}^R_{-2}(M')$$
$$\lambda : \mathbf{F} \longrightarrow \mathbf{F}' .$$

Moving to $\widetilde{R}$, we have

$$M(c-1) = \operatorname{Coker} \tilde{d}(c-1) = \mathrm{Syz}^{\widetilde{R}}_2(\mathrm{Syz}^R_{-2}(M))$$

and

$$M'(c-1) = \operatorname{Coker} \tilde{d}'(c-1) = \mathrm{Syz}^{\widetilde{R}}_2(\mathrm{Syz}^R_{-2}(M')) .$$

We will use the notation and the construction in the proof of Theorem 6.4.2, where we produced an $\widetilde{R}$-free resolution $\mathbf{V}$ of $\mathrm{Syz}^R_{-2}(M)$, and various homotopies on it. Of course we have a similar resolution $\mathbf{V}'$ of $\mathrm{Syz}^R_{-2}(M)$. See the diagram:

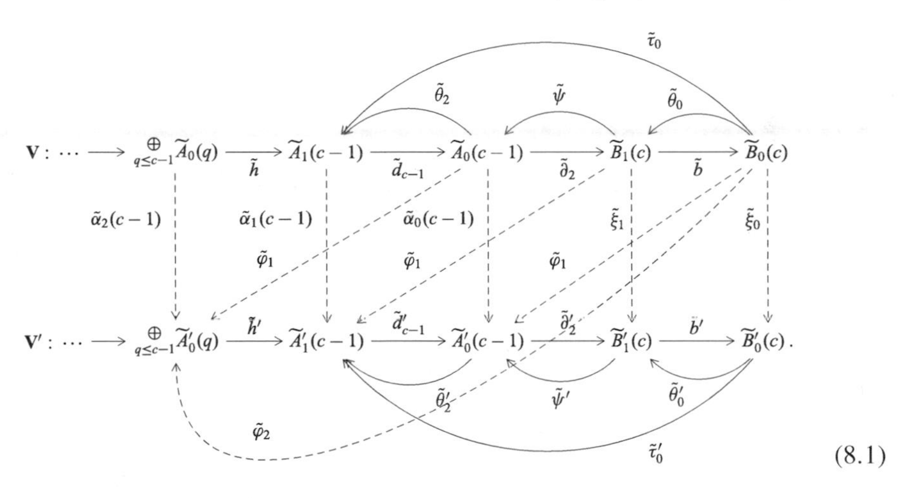

(8.1)

The map η induces $\tilde{\xi} : \mathbf{V} \longrightarrow \mathbf{V}'$, which in turn induces a map

$$\zeta(c-1) : M(c-1) \longrightarrow M'(c-1) .$$

See diagram (8.1).

By induction, the map $\zeta(c-1)$ is induced by an HMF morphism with components

$$\alpha_s(c-1) : A_s(c-1) \longrightarrow A'_s(c-1)$$

for $s = 0, 1$ and

$$\alpha_2(c-1) : \bigoplus_{q \leq c-1} \widetilde{A}_0(q) \longrightarrow \bigoplus_{q \leq c-1} \widetilde{A}_0(q)' .$$

By the conditions in 8.1.1, it follows that the first two squares on the left are commutative; clearly, the last square on the right is commutative as well. Since $\tilde{\alpha}_0(c-1)$ induces the same map on $M(c-1)$ as $\tilde{\xi}$, the remaining square commutes. Therefore we can apply Lemma 8.1.3 and conclude that there exists a system of homotopy comparison maps, the first few of which are shown as $\tilde{\varphi}_1$ and $\tilde{\varphi}_2$ in diagram (8.1).

With notation as in (6.13) and (6.14) we may write the first two steps of the minimal $\widetilde{R}$-free resolution $\mathbf{U}$ of M in the form given by the top three rows of diagram (8.2), where we have used a splitting

$$A_0(c) = A_0(c-1) \oplus B_0(c)$$

to split the left-hand term $\oplus_{q \leq c} A_0(q)$ into three parts, $\oplus_{q \leq c-1} A_0(q)$, $A_0(c-1)$, and $B_0(c)$; and similarly for M' and the bottom three rows. Straightforward computations using the definition of the system of homotopy comparison maps shows that diagram (8.2) commutes:

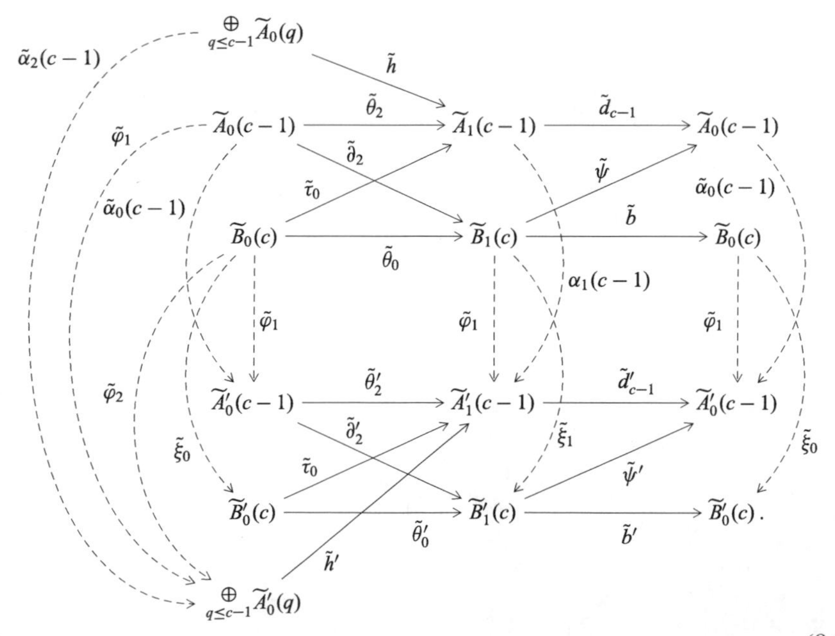

(8.2)

Next, we will construct the maps α_i. We construct α_0 by extending the map $\alpha_0(c-1)$ already defined over S by taking $\alpha_0|_{B_0(c)}$ to have as components arbitrary liftings to S of $\tilde{\xi}_0$ and $\tilde{\varphi}_1$. Similarly we take α_1 to be the extension of $\alpha_1(c-1)$ that has arbitrary liftings of $\tilde{\xi}_1$ and $\tilde{\varphi}_1$ as components. Finally, we take α_2 to agree with $\alpha_2(c-1)$ on $\oplus_{q\le c-1}A_0(q)$ and on the summand $A_0(c) = A_0(c-1) \oplus B_0(c)$, to be the map given by

$$\alpha_0(c-1) : A_0(c-1) \longrightarrow A'_0(c-1)$$

and arbitrary liftings

$$\varphi_1 : A_0(c-1) \longrightarrow \oplus_{q\le c-1}A'_0(q) \qquad \varphi_1 : B_0(c) \longrightarrow A'_0(c-1)$$

$$\xi_0 : B_0(c) \longrightarrow B'_0(c) \qquad \varphi_2 : B_0(c) \longrightarrow \oplus_{q\le c-1} A'_0(q)$$

to S of $\tilde{\varphi}_1, \tilde{\varphi}_2$ and $\tilde{\xi}_0$.

It remains to show that $\overline{\alpha}_0 = R \otimes_S \alpha_0$ induces $\zeta : M \longrightarrow M'$.

By Corollary 7.2.2 the minimal R-free resolutions $\mathbf{F}$ and $\mathbf{F}'$ of $\mathrm{Syz}^R_{-2}(M)$ and $\mathrm{Syz}^R_{-2}(M')$ have the form given in the following diagram:

$$\begin{array}{ccccccc}
\mathbf{F}: & \cdots \longrightarrow & \overline{B}_0(c) \oplus \overline{A}_0(c-1) & \xrightarrow{(\overline{\theta}_0, \overline{\partial}_2)} & \overline{B}_1(c) & \xrightarrow{\overline{b}} & \overline{B}_0(c) \\
 & & \overline{\alpha}_0 \Big\Downarrow \Big\downarrow \lambda_2 & & \Big\downarrow \lambda_1 = \overline{\xi}_1 & & \Big\downarrow \lambda_0 = \overline{\xi}_0 \\
\mathbf{F}': & \cdots \longrightarrow & \overline{B}'_0(c) \oplus \overline{A}'_0(c-1) & \xrightarrow[(\overline{\theta}'_0, \overline{\partial}'_2)]{} & \overline{B}'_1(c) & \xrightarrow[\overline{b}']{} & \overline{B}'_0(c),
\end{array}$$

By definition the map of complexes $\lambda : \mathbf{F} \longrightarrow \mathbf{F}'$ induces $\zeta : M \longrightarrow M'$. Using Lemma 8.1.3, we see that the left-hand square of the diagram also commutes if we replace λ_2 with α_0, and thus these two maps induce the same map $M \longrightarrow M'$, concluding the proof. □

References

1. P. Aspinwall, Some applications of commutative algebra to string theory, in *Commutative Algebra*, ed. by I. Peeva (Springer, Berlin, 2013), pp. 25–56
2. P. Aspinwall, D. Morrison, Quivers from matrix factorizations. Commun. Math. Phys. **313**, 607–633 (2012)
3. M. Auslander, R.-O. Buchweitz, The homological theory of maximal Cohen-Macaulay approximations. Mm. Soc. Math. France (NS) **38**, 5–37 (1989). Colloque en l'honneur de Pierre Samuel (Orsay, 1987)
4. L. Avramov, Modules of finite virtual projective dimension. Invent. Math. **96**, 71–101 (1989)
5. L. Avramov, Infinite free resolutions, in *Six Lectures on Commutative Algebra*. Modern Birkhauser Classics (Birkhäuser, Basel, 2010), pp. 1–118
6. L. Avramov, R.-O. Buchweitz, Homological algebra modulo a regular sequence with special attention to codimension two. J. Algebra **230**, 24–67 (2000)
7. L. Avramov, V. Gasharov, I. Peeva, Complete intersection dimension. Publ. Math. IHES **86**, 67–114 (1997)
8. L. Avramov, L.-C. Sun, Cohomology operators defined by a deformation, J. Algebra **204**, 684–710 (1998)
9. J. Backelin, J. Herzog, B. Ulrich, Linear maximal Cohen-Macaulay modules over strict complete intersections. J. Pure Appl. Algebra **71**, 187–202 (1991)
10. M. Ballard, D. Deliu, D. Favero, M.U. Isik, L. Katzarkov, Resolutions in factorization categories (2014). arXiv:1212.3264. http://arxiv.org/abs/1212.3264
11. M. Ballard, D. Favero, L. Katzarkov, A category of kernels for graded matrix factorizations and its implications for Hodge theory. Publ. Math. l'IHES **120**, 1–111 (2014)
12. R.-O. Buchweitz, G.-M. Greuel, F.-O. Schreyer, Cohen-Macaulay modules on hypersurface singularities II. Invent. Math. **88**, 165–182 (1987)
13. J. Burke, Complete intersection rings and Koszul duality (in preparation)
14. J. Burke, M. Walker, Matrix factorizations in higher codimension. Trans. Am. Math. Soc. **367**, 3323–3370 (2015)
15. M. Casanellas, R. Hartshorne, ACM bundles on cubic surfaces. J. Eur. Math. Soc. **13**, 709–731 (2011)
16. A. Cayley, On the theory of elimination. Camb. Dublin Math. J. **3**, 116–120 (1848). Collected papers: vol. I (Cambridge University Press, Cambridge, 1889), pp. 370–374
17. C. Curto, D. Morrison, Threefold flops via matrix factorization. J. Algebraic Geom. **22**, 599–627 (2013)
18. H. Dao, C. Huneke, Vanishing of ext, cluster tilting and finite global dimension of endomorphisms of rings. Am. J. Math. **135**, 561–578 (2013)

D. Eisenbud, I. Peeva, *Minimal Free Resolutions over Complete Intersections*, Lecture Notes in Mathematics 2152, DOI 10.1007/978-3-319-26437-0

19. W. Decker, G.-M. Greuel, G. Pfister, H. Schönemann, SINGULAR 4-0-2 computer algebra system for polynomial computations (2015), http://www.singular.uni-kl.de
20. S. Ding, Cohen-Macaulay approximations over a Gorenstein local ring. Ph.D. thesis, Brandeis University, 1990
21. T. Dyckerhoff, Compact generators in categories of matrix factorizations. Duke Math. J. **159**, 223–274 (2011)
22. T. Dyckerhoff, D. Murfet, Pushing forward matrix factorizations. Duke Math. J. **162**, 1249–1311 (2013)
23. A. Efimov, L. Positselski, Coherent analogues of matrix factorizations and relative singularity categories. Alg. and Number Theory **9**, 1159–1292 (2015)
24. D. Eisenbud, Enriched free resolutions and change of rings, in *Séminaire d'Algèbre Paul Dubreil* (Paris, 1975–1976). Lecture Notes in Mathematics, vol. 586 (Springer, Berlin, 1977), pp. 1–8
25. D. Eisenbud, Homological algebra on a complete intersection, with an application to group representations. Trans. Am. Math. Soc. **260**, 35–64 (1980)
26. D. Eisenbud, *Commutative Algebra. With a View Toward Algebraic Geometry*. Graduate Texts in Mathematics, vol. 150 (Springer, Berlin, 1995)
27. D. Eisenbud, I. Peeva, Standard decompositions in generic coordinates. J. Commut. Algebra (Special volume in honor of Jürgen Herzog) **5**, 171–178 (2013)
28. D. Eisenbud, I. Peeva, F.-O. Schreyer, Tor as a module over an exterior algebra (preprint), 2016
29. D. Eisenbud, I. Peeva, F.-O. Schreyer, Quadratic complete intersections (preprint), 2016
30. D. Grayson, M. Stillman, Macaulay2 – a system for computation in algebraic geometry and commutative algebra, http://www.math.uiuc.edu/Macaulay2/, 1992
31. T. Gulliksen, A change of ring theorem with applications to Poincaré series and intersection multiplicity. Math. Scand. **34**, 167–183 (1974)
32. D. Hilbert, Über die Theorie der algebraischen Formen. Maht. Ann. **36**, 473–534 (1890); Ges. Abh., vol. II (Springer, Berlin, 1933 and 1970), pp. 199–257
33. M. Hochster, The dimension of an intersection in an ambient hypersurface, in *Algebraic Geometry*. Lecture Notes in Mathematics, vol. 862 (Springer, Berlin, 1981), pp. 93–106
34. C. Huneke, R. Wiegand, Tensor products of modules and the rigidity of Tor. Math. Ann. **299**, 449–476 (1994)
35. M.U. Isik, Equivalence of the derived category of a variety with a singularity category (2010). arXiv:1011.1484. http://arxiv.org/abs/1011.1484
36. H. Kajiura, K. Saito, A. Takahashi, Matrix factorization and representations of quivers. II. Type ADE case. Adv. Math. **211**, 327–362 (2007)
37. A. Kapustin, Y. Li, D-branes in Landau-Ginzburg models and algebraic geometry. J. High Energy Phys. **12**, 1–43 (2003)
38. M. Khovanov, L. Rozansky, Matrix factorizations and link homology. Fund. Math. **199**, 1–91 (2008)
39. M. Khovanov, L. Rozansky, Matrix factorizations and link homology II. Geom. Topol. **12**, 1387–1425 (2008)
40. H. Knörrer, Cohen-Macaulay modules on hypersurface singularities. Invent. Math. **88**, 153–164 (1987)
41. J. McCullough, I. Peeva, Infinite free resolutions, in *Commutative Algebra and Noncommutative Algebraic Geometry*, ed. by Eisenbud, Iyengar, Singh, Stafford, Van den Bergh (Cambridge University Press, Cambridge), pp. 101–143
42. V. Mehta, Endomorphisms of complexes and modules over Golod rings. Ph.D. thesis, University of California at Berkeley, 1976
43. D. Orlov, Triangulated categories of singularities and D-branes in Landau-Ginzburg models. Tr. Mat. Inst. Steklova **246** (2004). Algebr. Geom. Metody, Svyazi i Prilozh, 240–262; translation in Proc. Steklov Inst. Math. **246**, 227–248 (2004)
44. D. Orlov, Triangulated categories of singularities, and equivalences between Landau-Ginzburg models (Russian. Russian summary). Mat. Sb. **197**, 117–132 (2006); translation in Sb. Math. **197**, 1827–1840 (2006)

45. D. Orlov, Landau-Ginzburg models, D-branes, and mirror symmetry. Mat. Contemp. **41**, 75–112 (2012)
46. D. Orlov, Derived categories of coherent sheaves and triangulated categories of singularities, in *Algebra, Arithmetic, and Geometry: In Honor of Yu.I. Manin*, vol. II. Progress in Mathematics, vol. 270 (Birkhäuser, Boston, MA, 2009), pp. 503–531
47. D. Orlov, Matrix factorizations for nonaffine LG-models. Math. Ann. **353**, 95–108 (2012)
48. I. Peeva, *Graded Syzygies* (Springer, Berlin, 2011)
49. G. Piepmeyer, M. Walker, A new proof of the New Intersection Theorem. J. Algebra **322**, 3366–3372 (2009)
50. A. Polishchuk, A. Vaintrob, Chern characters and Hirzebruch-Riemann-Roch formula for matrix factorizations. Duke Math. J. **161**, 1863–1926 (2012)
51. A. Polishchuk, A. Vaintrob, Matrix factorizations and cohomological field theories. arXiv:1105.2903 (2014)
52. F. Reid, Modular representations of elementary abelian p-groups (in preparation).
53. E. Segal, Equivalences between GIT quotients of Landau-Ginzburg B-models. Commun. Math. Phys. **304**, 411–432 (2011)
54. P. Seidel, Homological mirror symmetry for the genus two curve. J. Algebra Geom. **20**, 727–769 (2011)
55. J. Shamash, The Poincaré series of a local ring. J. Algebra **12**, 453–470 (1969)
56. I. Shipman, A geometric approach to Orlov's theorem. Compos. Math. **148**, 1365–1389 (2012)
57. J. Tate, Homology of Noetherian rings and local rings. Illinois J. Math. **1**, 14–27 (1957)
58. O. Zariski, The concept of a simple point of an abstract algebraic variety. Trans. Am. Math. Soc. **62**, 1–52 (1947)

Index

D. Eisenbud, I. Peeva, *Minimal Free Resolutions over Complete Intersections*,
Lecture Notes in Mathematics 2152, DOI 10.1007/978-3-319-26437-0

LECTURE NOTES IN MATHEMATICS Springer

Editorial Policy

1. Lecture Notes aim to report new developments in all areas of mathematics and their applications – quickly, informally and at a high level. Mathematical texts analysing new developments in modelling and numerical simulation are welcome.

   Manuscripts should be reasonably self-contained and rounded off. Thus they may, and often will, present not only results of the author but also related work by other people. They may be based on specialised lecture courses. Furthermore, the manuscripts should provide sufficient motivation, examples and applications. This clearly distinguishes Lecture Notes from journal articles or technical reports which normally are very concise. Articles intended for a journal but too long to be accepted by most journals, usually do not have this "lecture notes" character. For similar reasons it is unusual for doctoral theses to be accepted for the Lecture Notes series, though habilitation theses may be appropriate.

2. Besides monographs, multi-author manuscripts resulting from SUMMER SCHOOLS or similar INTENSIVE COURSES are welcome, provided their objective was held to present an active mathematical topic to an audience at the beginning or intermediate graduate level (a list of participants should be provided).

   The resulting manuscript should not be just a collection of course notes, but should require advance planning and coordination among the main lecturers. The subject matter should dictate the structure of the book. This structure should be motivated and explained in a scientific introduction, and the notation, references, index and formulation of results should be, if possible, unified by the editors. Each contribution should have an abstract and an introduction referring to the other contributions. In other words, more preparatory work must go into a multi-authored volume than simply assembling a disparate collection of papers, communicated at the event.

3. Manuscripts should be submitted either online at www.editorialmanager.com/lnm to Springer's mathematics editorial in Heidelberg, or electronically to one of the series editors. Authors should be aware that incomplete or insufficiently close-to-final manuscripts almost always result in longer refereeing times and nevertheless unclear referees' recommendations, making further refereeing of a final draft necessary. The strict minimum amount of material that will be considered should include a detailed outline describing the planned contents of each chapter, a bibliography and several sample chapters. Parallel submission of a manuscript to another publisher while under consideration for LNM is not acceptable and can lead to rejection.

4. In general, **monographs** will be sent out to at least 2 external referees for evaluation.

   A final decision to publish can be made only on the basis of the complete manuscript, however a refereeing process leading to a preliminary decision can be based on a pre-final or incomplete manuscript.

   Volume Editors of **multi-author works** are expected to arrange for the refereeing, to the usual scientific standards, of the individual contributions. If the resulting reports can be forwarded to the LNM Editorial Board, this is very helpful. If no reports are forwarded or if other questions remain unclear in respect of homogeneity etc, the series editors may wish to consult external referees for an overall evaluation of the volume.

5. Manuscripts should in general be submitted in English. Final manuscripts should contain at least 100 pages of mathematical text and should always include

   - a table of contents;
   - an informative introduction, with adequate motivation and perhaps some historical remarks: it should be accessible to a reader not intimately familiar with the topic treated;
   - a subject index: as a rule this is genuinely helpful for the reader.
   - For evaluation purposes, manuscripts should be submitted as pdf files.

6. Careful preparation of the manuscripts will help keep production time short besides ensuring satisfactory appearance of the finished book in print and online. After acceptance of the manuscript authors will be asked to prepare the final LaTeX source files (see LaTeX templates online: https://www.springer.com/gb/authors-editors/book-authors-editors/manuscriptpreparation/5636) plus the corresponding pdf- or zipped ps-file. The LaTeX source files are essential for producing the full-text online version of the book, see http://link.springer.com/bookseries/304 for the existing online volumes of LNM). The technical production of a Lecture Notes volume takes approximately 12 weeks. Additional instructions, if necessary, are available on request from lnm@springer.com.

7. Authors receive a total of 30 free copies of their volume and free access to their book on SpringerLink, but no royalties. They are entitled to a discount of 33.3 % on the price of Springer books purchased for their personal use, if ordering directly from Springer.

8. Commitment to publish is made by a *Publishing Agreement*; contributing authors of multiauthor books are requested to sign a *Consent to Publish form*. Springer-Verlag registers the copyright for each volume. Authors are free to reuse material contained in their LNM volumes in later publications: a brief written (or e-mail) request for formal permission is sufficient.

Addresses:

Professor Jean-Michel Morel, CMLA, École Normale Supérieure de Cachan, France
E-mail: moreljeanmichel@gmail.com

Professor Bernard Teissier, Equipe Géométrie et Dynamique,
Institut de Mathématiques de Jussieu – Paris Rive Gauche, Paris, France
E-mail: bernard.teissier@imj-prg.fr

Springer: Ute McCrory, Mathematics, Heidelberg, Germany,
E-mail: lnm@springer.com

Printed in the United States
By Bookmasters